Structural Engineering Documents
19

Seismic Isolation and Response Control

Andreas Lampropoulos
(Editor)

Authors (alphabetically)

Eftychia Apostolidi
Stephanos Dritsos
Christos Giarlelis
José Jara
Fatih Sutcu
Toru Takeuchi
Joe White

International Association for Bridge and Structural Engineering (IABSE)

ISBN: 978-3-85748-180-2 (print)
eISBN: 978-3-85748-179-6 (PDF), 978-3-85748-182-6 (ePUB)
DOI: https://doi.org/10.2749/sed019

Publisher

IABSE
Jungholzstrasse 28
8050 Zürich
Switzerland

Phone: Int. +41-43-443 9765
E-mail: secretariat@iabse.org
Web: www.iabse.org

This book was produced in cooperation with Structurae, Dresdener Str. 110, Berlin, Germany (https://structurae.net).

Copyediting: Jens Völker
Layout & typesetting: Florian Hawemann

Preface

The seismic resilience of new and existing structures is a key priority for the protection of human lives and the reduction of economic losses in earthquake-prone areas. The implementation of modern seismic codes for the design of new earthquake-resistant buildings and the advances in techniques for the repair and strengthening of existing deficient structures have focused on the upgrade of the structural performance of the new and existing structures. However, in many cases, it is preferable to mitigate the effects of earthquakes by reducing the induced loads in the structures using seismic isolation and response control devices. The main principle is that the use of appropriate seismic isolation and response control devices at the base of the structures will offer increased flexibility and energy absorption characteristics preventing resonance and significantly reducing the induced loads and deformations. The reduction of the deformations is also one of the main reasons for using these methods in cases of buildings with special requirements such as limited induced displacements in case of earthquakes (e.g. museums, hospitals, precision instruments and other equipment sensitive to displacements and accelerations etc.).

The use of seismic isolation and response control systems has become a quite popular technique not only for the design of new but also for the upgrade of existing structures. Various systems have been developed, and some limited information is also included in modern seismic codes for the design of new buildings with seismic isolation. However, the limited expertise on the selection of the appropriate system and its design for new and existing structures is the main challenge for practitioners and hinders the extensive use of seismic isolation and response control systems in practice. This is even more challenging for the application of these systems in existing structures where additional practical difficulties during the installation process are to be anticipated. The selection of the appropriate system depends on a large number of parameters, including the requirements and the particular characteristics of the examined structures. The engineers need to consider various possible systems, and the selection of the appropriate technology as well as the design process is in many cases a process with many iterations and alternatives.

The first part of this document is focused on the collection of the most commonly used seismic isolation and response control systems and the critical evaluation of the main characteristics of these systems. Then a comparison of the key parameters of the design processes for the design of new buildings with seismic isolation is presented, followed by four case studies from New Zealand, Greece, and Mexico and one case study on response control systems from Japan. The application of seismic isolation systems and response control systems for the retrofitting of existing structures were also examined. Two case studies on the application of seismic isolation systems in Turkey and Greece are presented, followed by three case studies on the application of response control systems in existing structures in Japan, Turkey and New Zealand. Finally, post-earthquake survey observations from seismic isolated structures are described to evaluate the efficiency of the application of these systems.

The main aim of this document is to provide a practical guide for the selection of seismic isolation and response control systems and explain the main steps of the design and application process. This work has been conducted as one of the main tasks of IABSE Task Group 1.1 'Improving Seismic Resilience of Reinforced Concrete Structures', which is part of Commission 1 'Performance and Requirements'.

This work was coordinated by the IABSE Task Group 1.1 Chairman Dr Andreas Lampropoulos (Editor) and presents a teamwork of the following members (listed in alphabetical order): Dr Eftychia Apostolidi, Professor Stephanos Dritsos, Mr Christos Giarlelis, Professor Jose Jara. Professor Fatih Sutcu, Professor Toru Takeuchi and Dr Joe White.

Chapter 1 (Introduction) was led by Professor Stephanos Dritsos and Chapters 2 (Seismic Isolation and Response Control Systems), 3.1 (Design of New Buildings with Seismic Isolation), and 3.2 (Basics of Seismic Isolation Design) were led by Professor Fatih Sutcu, Professor Toru Takeuchi and Mr Christos Giarlelis. Mr Christos Giarlelis also led the preparation of the three case studies in Greece. Professor Jose Jara led the preparation of the case study in Mexico. Professor Fatih Sutcu led the preparation of the two case studies in Turkey. Professor Toru Takeuchi led the preparation of the two case studies in Japan. Dr Joe White led the preparation of the two case studies in New Zealand. Dr Eftychia Apostolidi worked on the enhancement and completeness of the main part of the document. All the authors of the list contributed to various sections of this document which represents the outcome of a collective effort.

The Editor would like to express his appreciation and sincere thanks to the reviewers, Prof. Fabrizio Palmisano (Chief Reviewer, Editorial Board), Prof. Alberto Pavese and Asst. Prof. Bahadir Sadan, for their comprehensive and valuable comments and suggestions.

Finally, the Editor would like to express his gratitude to the Chair of IABSE Commission 1, Mr Niels Peter Hoj, and the Chair of the Bulletin Board, Dr Harsha Subbarao, for their continuous encouragement and support during the preparation of this document.

Dr Andreas Lampropoulos
(Editor)

Table of Contents

ANTI-SEISMIC TECHNOLOGY

UBB™ (Unbonded Brace™)

The most widely used BRB in the world

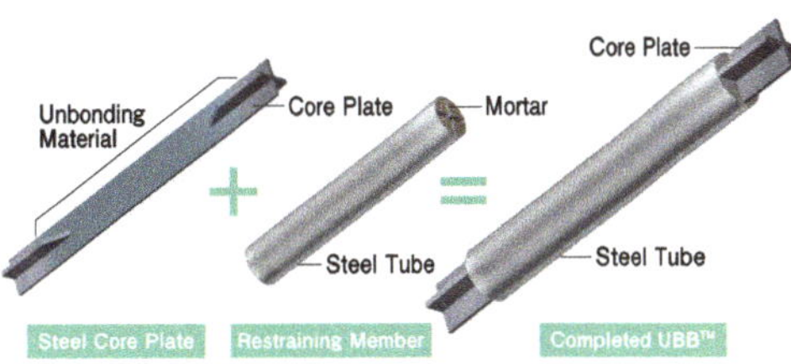

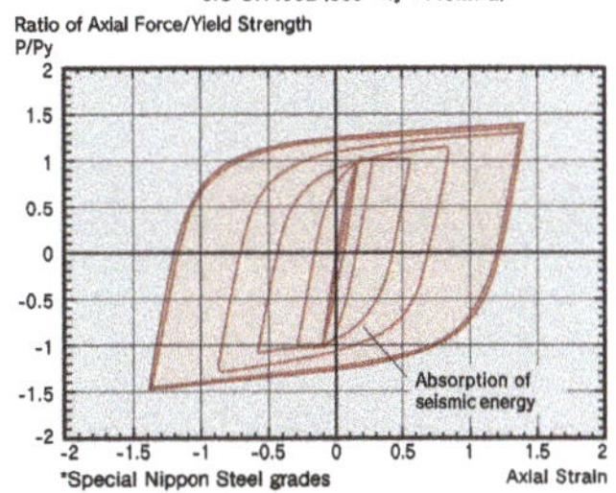

■Qualified to ANSI/AISC 341

Extensive range of UBBs™ with welded, bolted, pinned and custom connections have been qualified through physical testing to ANSI/ AISC 341 of the American Institute of Steel Construction (AISC).

SEISMIC ISOLATION TECHNOLOGY

NS-U™ (U-shaped Steel Damper™)

The ingeniously crafted steel damper

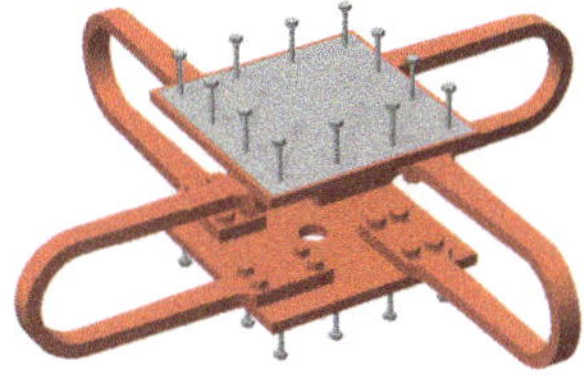

Static Loading Test (NSUD45×4)

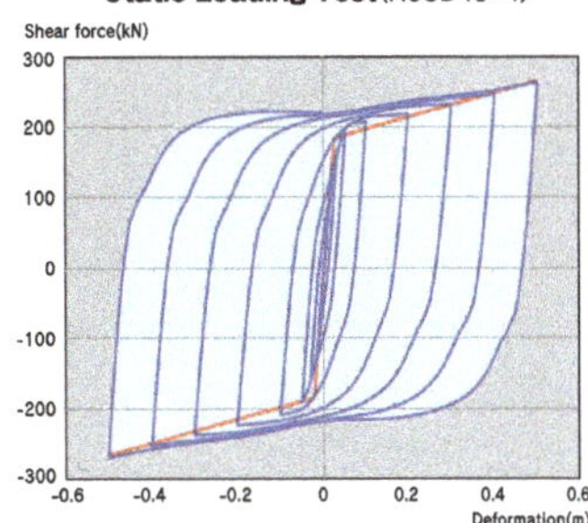

■Added damping for seismic isolation

Stable, bi-directional yielding behavior with large deformation and excellent fatigue capacities. Available as discrete units or integrated with natural rubber bearings.

ntacts
d office   OSAKI CENTER BUILDING, 5-1, Osaki 1-Chome, Shinagawa-ku, Tokyo 141-8604, Japan Tel.+81-3-6665-4330
           www.eng.nipponsteel.com/english/whatwedo/building/
nail   NSENGI_steel_structures@eng.nipponsteel.com

NIPPON STEEL ENGINEERING CO., LTD.

List of Abbreviations

AIJ	Architectural Institute of Japan
BRB	Buckling-Restrained Braces
CFD	Computational Fluid Dynamics
DBE	Design Basis Earthquake
DCLS	Damage Control Limit State
EEI	Environmental Energy Innovation
ELFM	Equivalent Lateral Force Method
FPS	Friction Pendulum System
FVD	Fluid Viscous Dampers
HDRB	High Damping Rubber Bearings
HVAC	Heating, Ventilation and Air Conditioning
IBC	International Building Code
LDD	Low Damage Design
LDRB	Low Damping Rubber Bearings
LRB	Lead-plug Rubber Bearings
MCE	Maximum Considered Earthquake
MEP	Mechanical, Electrical and Plumbing
NLTHA	Non-Linear Time-History Analysis
NRB	Natural Rubber Bearings
PGA	Peak Ground Acceleration
PRB	Polymer Plug Rubber Bearings
RC	Reinforced Concrete
RSA	Response Spectrum Analysis
SB	Sliding Bearings
SDOF	Single Degree of Freedom
SNFCC	Stavros Niarchos Foundation Cultural Centre
SSI	Soil-Structure Interaction
ULS	Ultimate Limit State

Chapter 1

Introduction

Traditionally, the design of structures has steadily followed the safety verification rule that, in any element of the structure, the design action effects should be lower than the respective resistance. Until now, the above verification is mainly performed in terms of forces. This so-called Force-Based Design has been the main design procedure adopted in our codes. Based on this approach, the safety of the whole structure can only be ensured when safety verification criteria are satisfied for all the elements of the structure without investigating the performance of the structure when the capacity of one or more structural elements is exceeded.

In the last 25 years, the engineering community has adopted the concept of displacement-based design, in which the safety verification is performed in terms of displacements not only of the members but also of the structure as a whole. Moreover, since the displacements are determined, the functionality of the structure can be verified as well.

Recently, the idea to integrate verification for safety, integrity, stability and functionality of a structure in a holistic way, through global verification criteria, for a set of earthquake scenarios has gained more and more ground/attention. The contribution of each element to the whole performance of the structure is considered, but, in general, there is no need to verify the integrity of each one of them, accepting a level of damage depending on the importance of the examined structure. This has been now introduced in the codes, using specific performance or damage levels where the relevant losses are addressed to the structure as a whole. Displacement-based design is a breakthrough approach for seismic engineering.

However, the increased needs of the modern overdeveloped and overpopulated societies and the alarming consequences of extreme events (e.g. earthquakes) can lead to a significant amount of fatalities and depleting resources and the subsequent collapse of the society. Therefore, there is an urgent need for the development of a 'resilient'-based approach.

A key factor for the development of this approach is the consideration and quantification of a wide range of direct and indirect losses. The engineering community is traditionally focused mostly on the repair and reconstruction costs after seismic events. However, losses of human life, monuments, historic structures, or exhibits in museums may have more value than the repair and reconstruction cost of damaged structures. In addition to these 'direct' consequences, indirect losses should be considered. These include not only the economic activity losses due to the inability of the people to continue their jobs or to use their houses, but also the diminishment of the quality of their lives.

The quantification and evaluation of the consequences of earthquakes and other hazards is a quite complex process, but, typically, their impact on the communities lasts for years and even decades after an event [1], [2].

Therefore, the pace and the total time of recovery after a seismic event plays a very significant role in the determination of the actual cost of losses. In Fig. 1.1, the interruption of the normal functionality of a structure after a strong seismic event is qualitatively demonstrated. The total recovery time, otherwise called downtime, is given by Eq. (1.1):

$$\Delta T = T_r - T_0 \tag{1.1}$$

where, T_o is the time when the earthquake strikes,

T_r is the time required for the structure to return to its previous functionality.

Obviously, minimum losses at the time of a seismic event, followed by a short recovery time, would be the most desired result. If the above concept is adopted in our seismic design, a Resilience-Based Design can be achieved.

Although many definitions of resilience can be found in the literature, the backbone of the term includes two crucial parameters, which are robustness and recovery. In the following, robustness is considered as the ability of any asset or entity (system, society, city, business, individual, structure) to maintain critical operations and functions in the face of a crisis [3]. A crisis can be defined as any natural or manmade threat (earthquake, flood, wind, blast or explosion, fire, chemical, biological or radiological agent), including also, an economic crisis or a pandemic. Recovery is the ability of any asset or entity to return and/or continue normal operations as quickly and effectively as possible after a disruption [3].

Therefore, the goal of Resilience-Based Seismic Design of minimum direct and indirect losses is achieved through high robustness and quick recovery. With regards to Figure 1.1, this can be depicted by the minimum possible dashed area that corresponds to the loss of resilience.

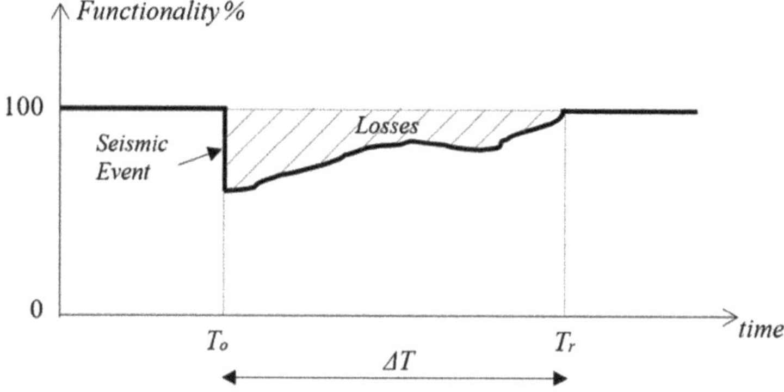

Fig. 1.1 Qualitative demonstration of functionality after a seismic event

In the above design framework, seismic isolation and response control systems can be considered as the most appropriate methods to achieve high resilience of a structure after a strong earthquake. These applications reduce the demands on both the structural and non-structural elements of the building, offering high robustness and quick recovery by preventing permanent damages to the structure and installing systems that can 'easily' be replaced in case of extreme events. These systems will lead to significantly reduced damages and subsequent reduction of direct and indirect losses. Repair/reconstruction costs and financial losses were reported [2] to be significantly lower than the respective losses following the usual code design. Moreover, seismic isolation and response control systems seem to be the most appropriate methods to design or retrofit structures where possible damage could result in irreplaceable losses, such as cultural or extensive life losses. Consequently, the above high resilient methods should be in the first line of choices in case of monumental or historical structures, museums and nuclear reactors, when conventional methods cannot offer the required seismic protection. Also, the methods are very suitable for hospitals, fire stations, power plants, water treatment plants, important administration buildings, police stations or structures of telecommunication and broadcasting services, since the interruption of their services, during and after a strong earthquake, should considerably reduce the overall society resilience. Another important advantage of the above methods is that they fulfil the "design for all" rule, since vulnerable groups of people, such as disabled people, young children and the elderly, should be treated with the same safety level as any other

As far as the cost of the above techniques is concerned, in many cases, it was found that the above methods offer high resilience at a minimal additional investment. It is worth mentioning that the application of seismic isolation in some cases resulted in quite lower construction costs (in the order of 30%) when compared to alternative design solutions

(a)

(b)

Fig. 1.2 (a) Oakland City Hall (Photo: Sanfranman59, CC-BY-SA 3-0 [5]) and (b) Pasadena City Hall (Photo: Bobak Ha'Eri, CC-BY-SA 3.0 [6])

without isolation [4]. However, one should also consider the additional costs concerning maintenance and potential replacement of the isolation system.

Until now, thousands of important structures around the world have been designed or retrofitted using seismic isolation or response control systems. It is worth mentioning some examples of high cultural value such as the Salt Lake City and County Building in Utah (the 1st application of seismic isolation in an existing building in the USA), the Oakland City Hall (Figure 1.2(a)), the Pasadena City Hall (Figure 1.2(b)), the James R. Browning U.S. Court of Appeals Building, the National Museum of Western Art in Tokyo (the 1st application of seismic isolation of an existing building in Japan) and the New Zealand Parliament Library and Parliament House in Wellington (the 1st application of seismic isolation in an existing building in New Zealand).

Chapter 2

Seismic Isolation and Response Control Systems

Seismic isolation and response control are seismic protection methods that are used to protect structures, non-structural components and contents of buildings from the damaging effects of earthquakes.

Seismic isolation is a method that is implemented to shift the natural vibration period of a structure to the long period range of approximately 2.0~4.0 secs by placing isolation bearings usually at the foundation level in order to physically decouple the structure from the ground. However, there are exceptions where seismic isolation may be used in the upper floors of a structure or that the fundamental period of the isolated structure exceeds 4.0 secs. The isolation layer consists of horizontally flexible devices that are capable of reducing the lateral stiffness of the superstructure combining structure re-centring and energy dissipation capability. Energy dissipation in the bearings (or separate dampers placed in parallel) can increase the effective damping ratio of the whole system. The result is a reduction of the acceleration and shear force response. This approach is most suitable for low-to-mid-rise stiff structures where a clear separation between the natural period of the flexible isolator bearings and a stiff superstructure minimises the transfer of lateral accelerations. In these cases, typically, a seismic isolation system can be designed that enables the superstructure to remain elastic even after a Maximum Considered Earthquake (MCE) event.

Response control, on the other hand, is a technique where structures are equipped with energy dissipating devices that will improve the structural integrity, reduce the dynamic responses of the structures or enable the control of higher mode effects during dynamic excitations such as seismic events or winds.

With their given merits, these methods provide the highest possible seismic protection for building structures. Often, equally important, damage to non-structural items (partitions, ceilings, façades and building services etc.) can be prevented or significantly reduced. Furthermore, the contents of the structures are better protected since the induced accelerations are lower. When a building and/or its contents are of high importance (such as hospitals, data centres, transportation facilities, etc.), seismic isolation and response control techniques enable the buildings to remain functional even after a major earthquake. Similarly, these techniques can be implemented in industrial buildings, bridge structures or other important facilities, where uninterrupted operation is required, during and after an earthquake. These methods can be applied to new structures as well as be used as a retrofit solution in existing structures.

Both methods can be considered innovative applications in earthquake engineering, and the number of applications increases day by day worldwide.

2.1 Basic Concepts

The seismic response of a structure depends on its natural period and damping ratio. Figure 2.1 shows a typical seismic spectrum according to various international standards. To develop a seismic isolation or response control concept in conventional strength-designed Reinforced Concrete (RC) structures, the following points have to be considered:

- Natural Period: Peak accelerations are induced in structures with a natural period in the approximate range of 0.2 ~ 0.6 secs that correspond to the horizontal spectrum branch of Figure 2.1. Accelerations reduce significantly at lower or higher periods in sites of stiff soil, as indicated by the ascending and descending branches of the spectrum (Figure 2.1). Low-rise buildings may have very short natural periods (i.e. less than 0.2 secs), but as the damage occurs in the initial stages of an earthquake, the natural period can be shifted to the 0.2~0.6 secs range, which in general is the natural period range of medium height buildings. Therefore, low to mid-rise buildings can significantly benefit from seismic isolation, although, even for high-rise buildings, seismic isolation is still beneficial.
- Damping Ratio: Conventional strength-designed RC structures generally have a low damping ratio, which may lead to significant damage, as has been shown in past earthquakes. Any increase in a structure's damping at most common structural periods reduces acceleration.

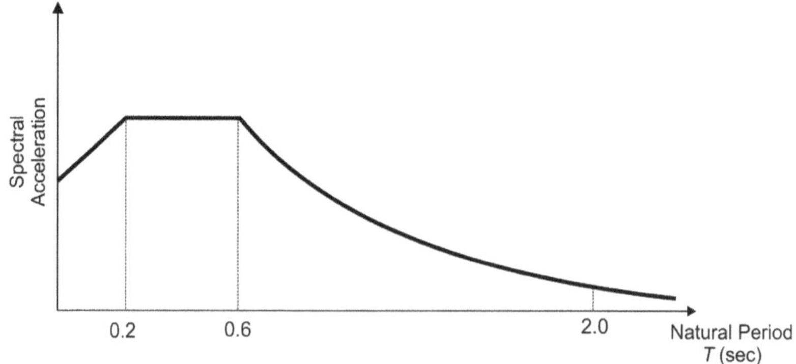

Fig. 2.1 Typical seismic spectrum according to various international standards

To improve these characteristics, two approaches have been developed over the past several decades, which are, as mentioned above, seismic isolated and response-controlled systems.

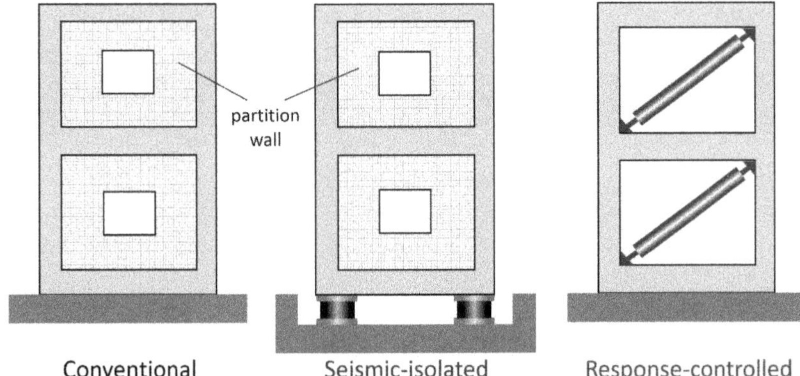

Fig. 2.2 Conceptual depiction of conventional, seismic isolated and response-controlled systems

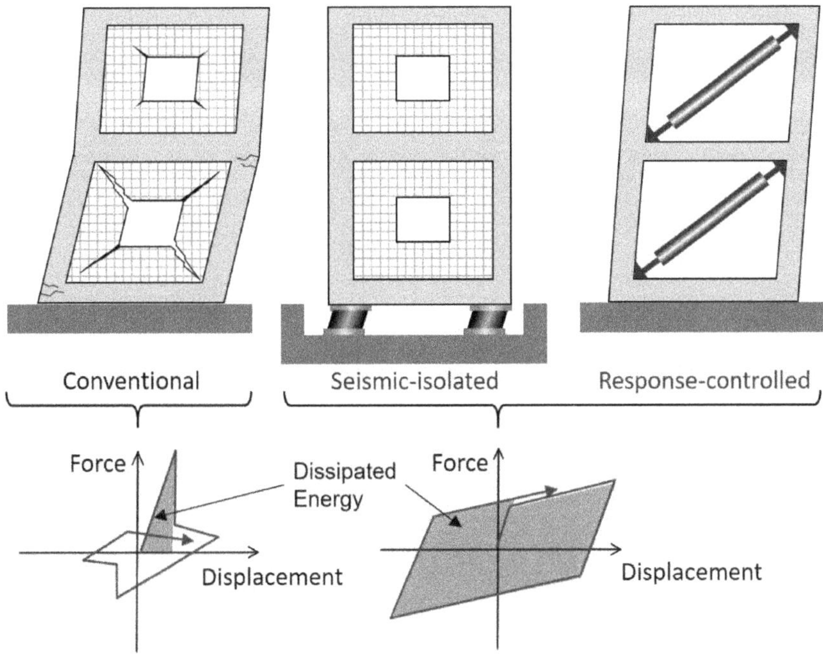

Fig. 2.3 Effect of seismic isolation and response-control systems on energy dissipation of a low-rise RC frame building during an earthquake

Figure 2.2 presents a conceptual low-rise RC frame before a seismic event in three different states; conventionally designed (fixed base), seismic isolated, and response controlled.

In Figure 2.3, the same RC frame in the respective aforementioned three states is presented, but this time during an earthquake. The conventionally built structure has undergone severe damage, while the seismic-isolated and response-controlled buildings remain almost undamaged, as the isolation and energy dissipating devices undertake the deformation introduced by the seismic loads.

Force-displacement relations during a seismic event for these three types of structural systems are depicted in Figure 2.3. Force-displacement curves show the changes in the structural behaviour and the enclosed area of cycles is the dissipated energy during the seismic motion. Seismic isolated and response-controlled systems have the ability to dissipate energy and undergo the deformations induced by the earthquake, as illustrated in Figure 2.3.

For taller and more flexible structures that have relatively longer natural periods, response-controlled concepts can be employed. Such systems reduce the response of structures by increasing the damping ratio with the insertion of energy-dissipation devices. If it is possible to keep the primary structure elastic and to limit spectral accelerations via energy dissipation in the dampers, then immediate occupancy after the earthquake can be achieved. If not, significant improvement in the seismic response can still be offered compared to a standard approach. In general, response-controlled systems prove to be more economical than seismic isolation systems.

The role of the soil should also be examined. Despite numerous studies that have been carried out over the years, there is still some controversy regarding the role of Soil-Structure Interaction (SSI) in the seismic performance of structures founded on soft soil [7].

In design practice and in the case of stiff soils, neglecting SSI effects has been considered as a conservative simplification that would facilitate analyses. However, in some cases and for specific (generally weak) soil types, there is a risk of settlements, and in this case, SSI should be taken into consideration. Regarding seismic isolated structures, the combined effect of SSI and seismic isolation has triggered the interest of researchers. The findings of some of these works are of particular interest. An experimental study conducted by Kelly [8] on the response of seismic isolated nuclear facilities resting on soft soil revealed that the design should consider significant displacement demands due to SSI. Tsai et al. [9] presented a study on the interactive behaviour of a Friction Pendulum System (FPS) isolated building and the soil and concluded that SSI results in larger displacements and, in some cases, also in larger shear forces. Spyrakos et al. [10] showed that SSI affects the modal properties of the system but has little effect on damping. More recently, Manolis and Markou [11] conducted a numerical SSI study examining the interplay of various parameters that affect the structural response of a seismically isolated structure and advocated the importance of SSI in fine-tuning the structure's design to the specific geological conditions and seismicity.

In all the aforementioned studies, the structures lie on shallow foundations. However, should a properly designed pile foundation be used, it would result in minimizing SSI effects [12]. In order to achieve good design, eliminating any relative displacements of the pile heads and increasing the stiffness of the soil-foundation system are considered key features. The first is achieved by creating a strong diaphragm connecting all pile heads, while optimisation of the number of piles is necessary in order to increase the stiffness.

In the following sections, detailed configurations of the most used devices of seismic isolation and response-controlled systems are described.

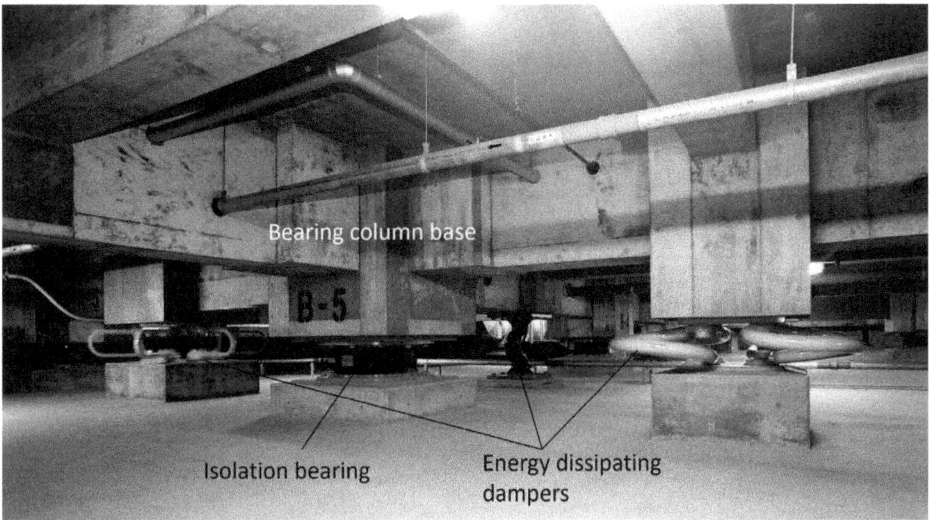

Fig. 2.4　Typical layout of devices in isolation layer in Japan (Photo: T. Takeuchi)

2.2　Seismic Isolation Systems

A seismic isolation system can consist of two components with respective functions as described below:

- Isolation components, which support gravity loads from the structure above and create an isolation plane, where significant movement can occur. This isolation plane increases the natural period of the structure.
- Energy-dissipation components, which increase the effective damping ratio.

In many countries (e.g. New Zealand, Greece, Italy, USA, Turkey), these two components and functions are typically combined into the same device. Beneath each column and load-bearing wall, a device is placed that contains an isolation bearing, which also provides damping.

In other countries (e.g. Japan), the isolation bearings and dampers are kept separate. Isolation bearings are placed under the bearing column/wall bases, and separate energy-dissipating dampers are placed below the foundation beams at locations such as at the mid-span since the dampers do not support vertical loads. A typical layout of devices in a Japanese system is shown in Figure 2.4.

2.2.1　Types of Seismic Isolation Bearings

In current practice, there are two major types of seismic isolation bearings that are also commonly included in the international codes, which are the rubber isolators (elastomeric seismic isolation bearings) and friction pendulum bearings (spherical sliding bearings). In addition to these two types, there are several novel devices that are successfully

Fig. 2.5 Spring-viscous damper system as installed under a three-storey town-house, Santa Monica, California (Photo: Shustov. CC-BY-SA 3.0 [14])

used in seismic isolation of building structures, as well as bridge structures, heavy energy or HVAC equipment or gas tanks. Many others are under the research and development process. A comparison table showing the advantages/disadvantages of several isolator types is given in sub-section 2.4.

As an example of novel devices, spring-type isolators can be counted (Figure 2.5). The main advantage of the system is that it is not affected by uplift. However, the fundamental period is generally not extended beyond 2.0 sec. The application of the system uses large helical steel springs that are flexible both horizontally and vertically. Therefore, this can work effectively in all three directions as opposed to the majority of other existing systems that provide seismic isolation only in the horizontal directions. Usually, this system works in conjunction with a damping device.

Three-dimensional isolation is another novel example in which natural rubber bearings, air springs, viscous dampers and vertical sliders are combined in order to isolate all components of seismic motion [13]. This unique application was used in Japan, Figure 2.6.

For the evaluation of the properties of seismic isolation devices, there are testing procedures that are categorised into two parts: prototype tests and production tests. Generally, two additional isolators are manufactured for the prototype destructive testing, which are disposed of after the end of the testing. Prototype tests are conducted in a project only if the required devices have unique dimensions and material properties. Prototype test results are used for the design of seismic isolated structures. Prototype tests are dynamic tests that should be conducted in independent testing centres, and the specimens are full scale. However, in some cases, prototype testing is conducted at the dynamic testing facility of the manufacturer, mostly supervised by an independent reviewer. Most prototype test loading protocols consist of design axial loads and several levels of lateral loads representing different levels of seismic effects. Production tests are "quality control tests", ensuring the performance of the seismic isolation bearings is in line with the design margins. Production tests are normally conducted in house by the manufacturer on actual bearings produced for a certain project. However, sometimes the client may require additional independent tests. Depending on the relevant code, production tests are conducted on 100 % of the bearings or on a percentage of the devices designated by the code, such as 30 % or higher. Production tests generally consist of 2 tests, an axial loading test for confirming the axial stiffness of the device and a lateral loading test with axial load to confirm the lateral stiffness and damping properties of the bearings.

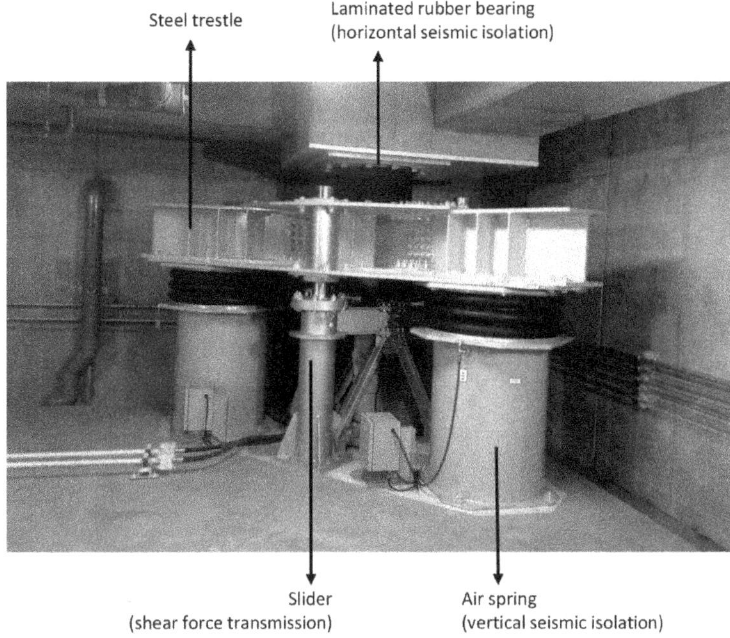

Fig. 2.6 Three-dimensional isolation system application (Photo: F. Sutcu)

2.2.1.1 Laminated Rubber Bearings

One of the most common types of isolation bearings is a laminated Natural Rubber Bearing (NRB), which is composed of rubber sheets and steel plates or fibre sheets. NRB is a popular choice for an isolator when natural rubber is available. However, synthetic rubber synthesised from petroleum is also extensively used. The great advantage in the latter case is the constancy of the mechanical characteristics and the possibility to calibrate them based on the components used and the production process. Laminated Natural Rubber Bearings are also suited to limit the instability of the device.

The components of a typical rubber bearing are shown in Figure 2.7. A full rubber block is flexible in the horizontal plane but also in the vertical plane and, therefore, is not suitable to support the weight of a building. By combining the laminated rubber with steel sheets, the vertical stiffness is dramatically increased to approximately 2,000 times stiffer than in the horizontal direction. This means that the bearing can support the weight of the superstructure whilst keeping a long natural period in the horizontal direction. Also, these bearings self-centre after an event.

A standard natural rubber bearing itself provides low damping. However, it is possible to add damping by using high-damping rubber or by inserting a lead core, which provides damping as it yields (Figure 2.8). Yielding of the lead core requires relatively large shear forces. Thus, these bearings are not suited to regions with low seismicity. Although uplift/tension may have some detrimental effect on the isolation systems, this is not covered in the document. Lead-plug Rubber Bearings (LRB) are one of the most popular devices in the USA and New Zealand, where there is significant seismicity, and combined bearings/

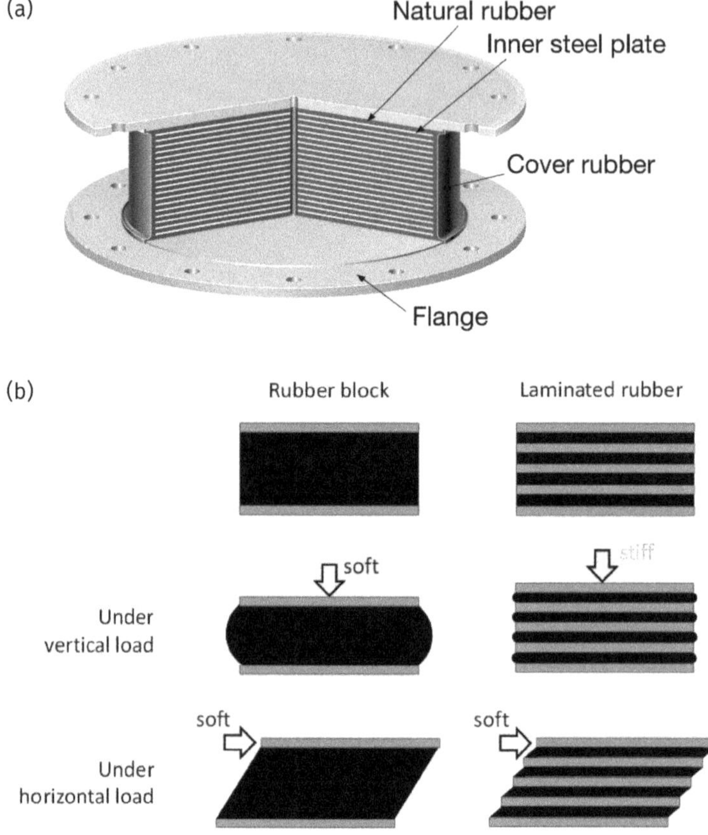

Fig. 2.7 Laminated Natural Rubber Bearing (NRB): (a) 3D view [15] and (b) deformation under loading conditions

dampers are preferred. On the other hand, High Damping Rubber Bearings (HDRB) look similar to NRBs in cross-section. It should be mentioned that specially developed rubber compounds with substantial energy dissipation capacity and HDRBs are widely used in Japan. NRBs usually provide low damping ratios (about 7 %), while HDRBs and LRBs can reach up to 20~30 %.

2.2.1.2 Friction Pendulum System (FPS) Bearings

Another commonly used device is the FPS bearing shown in Figure 2.9. This device also combines an isolation bearing and a damper. The device is composed of sets of concave plates with a single slider inserted between them for single and double pendulum isolators or more sliders for triple friction pendulum or pendulum with hinged sliders. When horizontally displaced, a slider moves along the spherical surfaces of the concave plates. The superstructure is thus isolated, moving like a pendulum along with the top plate of the device. Friction between the slider or sliders and the concave plates dissipates energy and, thus, provides damping. Also, these devices have usually rotational capabilities. The

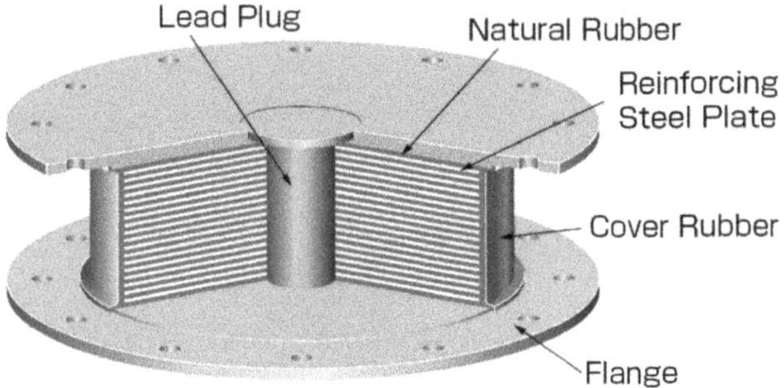

Fig. 2.8 Lead-plug Rubber Bearing (LRB) [15]

bearings' natural period is a function of the radius of curvature of the plates' spherical surface. Compared to laminated NRBs, the natural period of FPS bearings is easier to predict, especially with a lightweight structure or large displacements. The effective period of the isolation system is determined by the radius and friction properties of the bearing surface, so the behaviour of the isolation system is independent of the vertical load. Torsion motions of the structure are minimised because the centre of stiffness of the isolators coincides with the centre of mass of the structure, while it should also be mentioned that FPSs cannot bear tensile actions. Although uplift/tension may have some detrimental effect on the isolation systems, this is not covered in the document.

Single and double pendulum (Figure 2.9(a)) bearings are widely used. However, there are also triple pendulum bearings (Figure 2.9(b)). The latter are activated under lower seismic forces, which results in improved performance under earthquakes with lower intensity. Historical development, simplified design procedures, experimental dynamic response of FPS bearings, as well as advanced topics such as frictional heating of sliding surfaces, effects of mounting imperfections or restoring capability, are widely covered in the literature. With increasing research in the field, the characteristics and the performance of such devices are continuously improved, and at the same time, more reliable models are developed for the analysis and design of these systems. A series of research papers describe the state of the art on the design of these systems [16]–[20].

The equations for the design of the FPS are relatively simple. The structure can be modelled as an SDOF system of mass, m, based on an FPS type isolator. The behaviour of the system is described by the bilinear model of Figure 2.10 [21]–[23]. The effective stiffness, K_{eff}, is calculated using Eq. (2.1).

$$K_{eff} = \frac{N_{sd}}{R} + \frac{\mu\, N_{sd}}{D} \qquad (2.1)$$

Where:
    N_{sd} : normal load on the bearing, given by Eq. (2.2)
    R : radius of curvature of the sliding surface
    μ : friction coefficient (typical range of values 0.01–0.08)
    D : Horizontal Displacement of the System

(a)

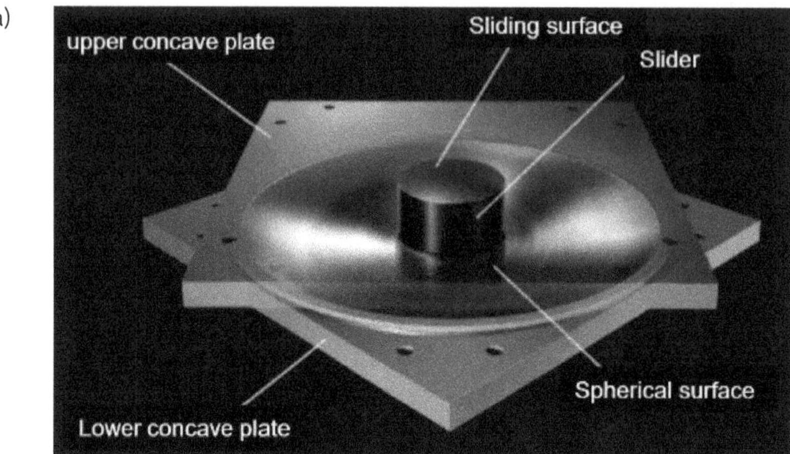

(b)

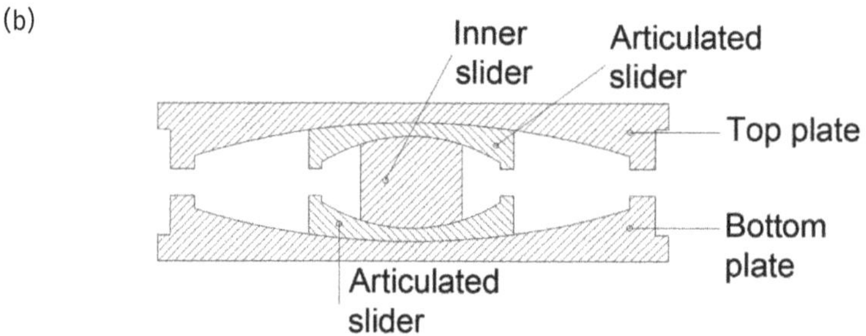

Fig. 2.9 (a) Double [24] and (b) triple friction pendulum bearings

$$N_{sd} = mg \qquad (2.2)$$

The first term of Eq. (2.1), K_h, shown in Figure 2.10, is related to the restoring force due to the rise of the mass, while the second term is related to the energy dissipation. The fundamental period of the system is given by Eq. (2.3).

$$T_{eff} = 2\pi \sqrt{\frac{m}{K_{eff}}} \qquad (2.3)$$

After substituting parameters K_{eff} and m from Eq. (2.1) and (2.2) respectively, Eq. (2.3) results to Eq. (2.4).

$$T_{eff} = 2\pi \sqrt{\frac{R\,D}{g\,D + \mu\,g\,R}} \qquad (2.4)$$

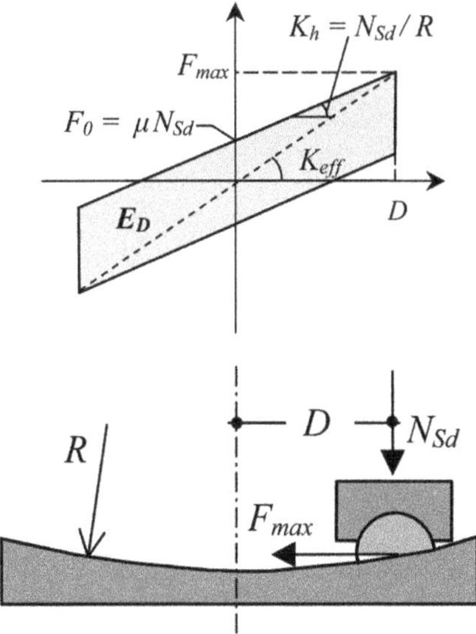

Fig. 2.10 Bilinear model of an FPS bearing

Thus, the effective period depends on the radius of curvature of the sliding surface, the friction coefficient, and the horizontal displacement of the system. For the calculation of damping coefficient, ξ, Eq. (2.5) can be used [23]:

$$\xi = \frac{1}{2\pi} \frac{\sum E_D}{K_{eff} D^2} \tag{2.5}$$

Where:

D : the displacement of the structure in the considered direction

E_D : the dissipated energy on a bearing in each cycle of displacement given by the area of the hysteresis loop of the bilinear model shown in Figure 2.10, which is given by the equation:

$$E_D = 4\mu N_{sd} D \tag{2.6}$$

For an SDOF system, Eq. (2.5) after substituting E_D from Eq. (2.6) reduces to:

$$\xi = \frac{2}{\pi} \frac{\mu}{D/R + \mu} \tag{2.7}$$

2.2.1.3 Flat Sliders

In some cases, in order to keep energy dissipation at a certain level and limit lateral stiffness and base shear, Sliding Bearings (SB) may be used (Figure 2.11). They are often used in combination with LRBs. For example, instead of placing LRBs under each column of a structure, thus, providing increased damping and stiffness, Flat Sliders could also be used in combination. If uplift loads must be resisted, two-way linear sliders (cross-linear

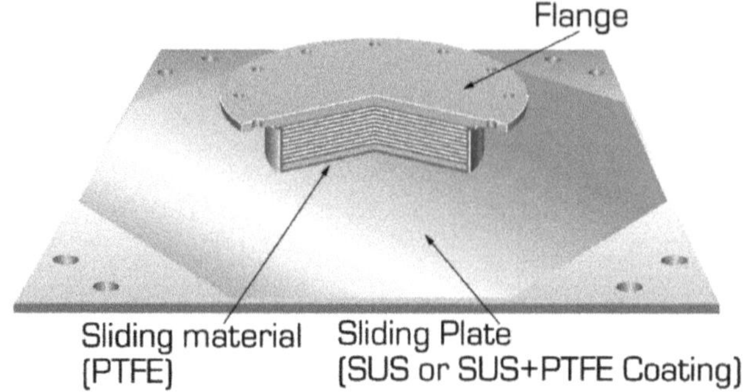

Fig. 2.11   Sliding bearing [15]

(a)   (b)

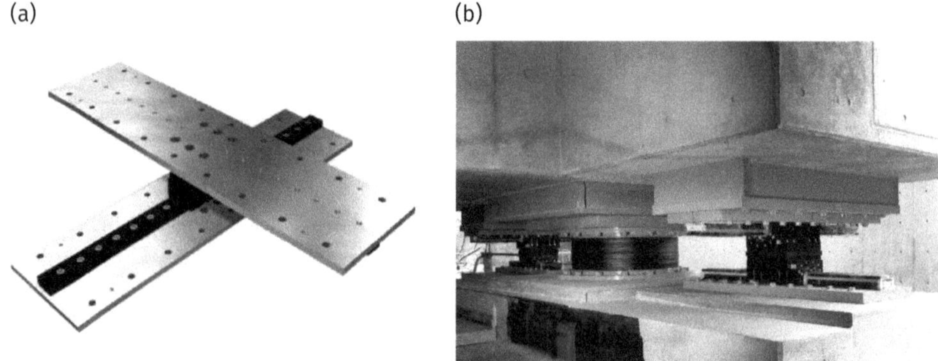

Fig. 2.12   (a) Linear slider [25] and (b) practical application in structure (Photo: T. Takeuchi)

bearings) can be used, as shown in Figure 2.12(a). Special care should be given to possible relative vertical displacements when a combination of isolators is used.

Flat sliders should not be used with FPS since due to the upward motion of FPS during their movement, the contact between the sliding surfaces of flat sliders would be lost. Being supplemental devices in a seismic isolation system, flat sliders do not have the capacity to re-centre. Re-centring is dependent on the main seismic isolation device such as LRBs. Therefore the ratio of sliders and LRBs in a building should be well proportioned, and their location should be carefully selected. An application of cross-linear bearing is shown in Figure 2.12(b).

2.2.2 Energy-Dissipating Components for Seismic Isolation

In countries with high seismic risk, such as Japan, where isolation bearings and dampers are kept separate in the seismic isolation layer, specific energy-dissipating components need to be introduced. One of the most widely used dampers is a U-shaped steel damper, as

(a)   (b)

(c)   (d)

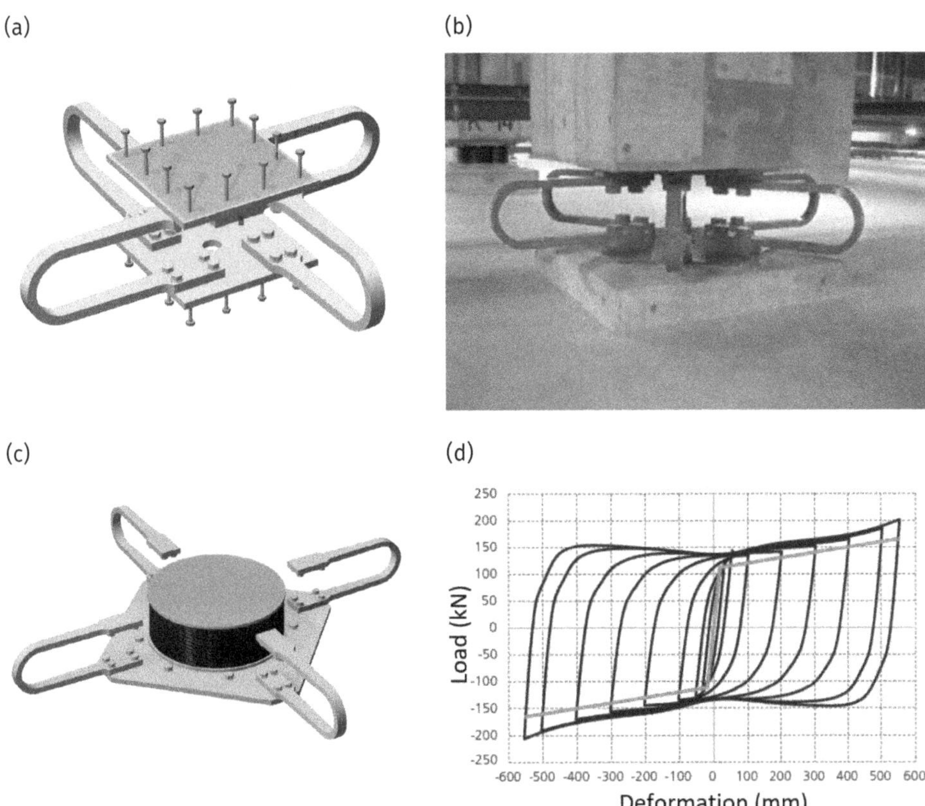

Fig. 2.13   (a) U-shaped steel damper and (b) its practical application in a column (Photo: S. Yamada) (c) U-shaped steel damper combined with a laminated rubber bearing. (d) Hysteresis of a U-shaped steel damper [26]

shown in Figure 2.13(a) and (b). This device can exhibit stable hysteresis up to ±500 mm of horizontal movement. They are often used in combination with Laminated NRBs (Figure 2.13(c)). Compared to LRBs, the inspection and replacement of damaged U-shaped steel damper components after an earthquake is easier. However, it should be mentioned that well-designed LRBs rarely need to be changed after a seismic event as opposed to U-shaped dampers. In Figure 2.13 (d), typical hysteresis loops of a U-shaped damper are shown, with the red line representing the backbone curve used for design purposes.

Two other types of seismic isolation dampers are the lead damper, shown in Figure 2.14, and the Fluid-Viscous (oil) Dampers (FVD), presented in Figure 2.15. The lead damper has a displacement-dependent hysteretic behaviour, whereas the viscous damper produces reaction forces as a function of velocity. To limit reaction forces developed in a viscous damper, a force limiter is often introduced to moderate the damping coefficient that is achieved at trigger velocities. Figure 2.14 shows images from a cyclic test of a lead damper. Here, several stages of a cyclic test are given; the initial condition (±0 mm), 400 mm displacement that corresponds to the design displacement, and 800 mm displacement that corresponds to the ultimate condition.

Fig. 2.14   Lead damper (Photo: T. Takeuchi) [27]

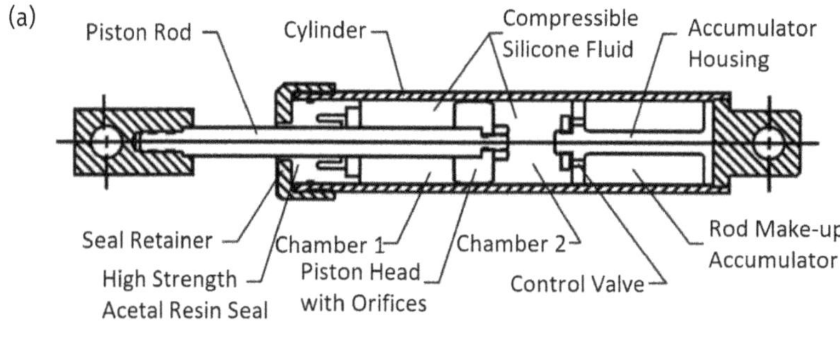

Fig. 2.15   Fluid-Viscous (oil) Dampers (FVD) (a) sketch of its main components [28] and (b) characteristic photograph (Photo: F. Sutcu)

Figure 2.15(a) and (b) depict the main components and a characteristic application of an oil viscous damper, respectively.

2.2.3 Selection of Seismic Isolation Systems

Selecting the most appropriate seismic isolation system depends on the project characteristics [12]. In this section, some key aspects for the selection of the most suitable system are summarised.

2.2.3.1 Rubber (Elastomeric) Bearings

The main advantage of the system is that, in many cases, the cost is lower compared to the other systems. However, this has been changing recently as mass production of other types of bearings lowers their cost.

The two principal disadvantages of elastomeric bearings are:

i) The behaviour of a base isolation system using elastomeric bearings varies with the vertical load applied to the bearing [29]. For most buildings, their self-weight is much larger than the transient live loads, so this effect is small. However, in certain types of buildings (cultural buildings, hospitals, etc.), the actual live load could be significant and variable.

ii) The properties of these bearings change over time due to ageing, which depends on the rubber compound and, generally, results in increased stiffness and characteristic strength. These increments are of the order of 10–20 % over a period of 30 years for low damping compounds but may be larger for materials compounded for high damping [29]. Thus, to remain effective, elastomeric bearings should be inspected and possibly periodically replaced.

2.2.3.2 Friction Pendulum System (FPS) Bearings

These bearings are well suited to uses where the live load contributes a large proportion of the vertical load. Friction between the sliding surfaces dissipates energy providing damping, so, in general, additional dampers are not required. Regarding their longevity, ageing is slow and limited [30]. However, the available information is not conclusive since field information is only available approximately for the last 25 years. Therefore, whilst the initial building cost of pendulum bearings may be higher than that of elastomeric ones, the former requires less or no periodic replacement due to ageing. Therefore, the total cost during the desired life cycle of the project can be lower.

2.2.4 Replacement of Bearings

As required by the codes, the bearings used in a project should pass a series of tests, which ensure their longevity and good performance in case of an earthquake. However, should they need replacement, the superstructures can be appropriately supported using jacks. The resulting stress condition during a possible replacement should be taken into account in the dimensioning and the reinforcement of the affected structural elements [12], [31]. It should be mentioned that this constitutes an additional structural study.

2.3 Response Control Systems

Structural dampers are energy-dissipation devices that are used for seismic mitigation of buildings. Various configurations of energy-dissipation devices for structural response control are shown in Figure 2.16. In many cases, these dampers are inserted into moment

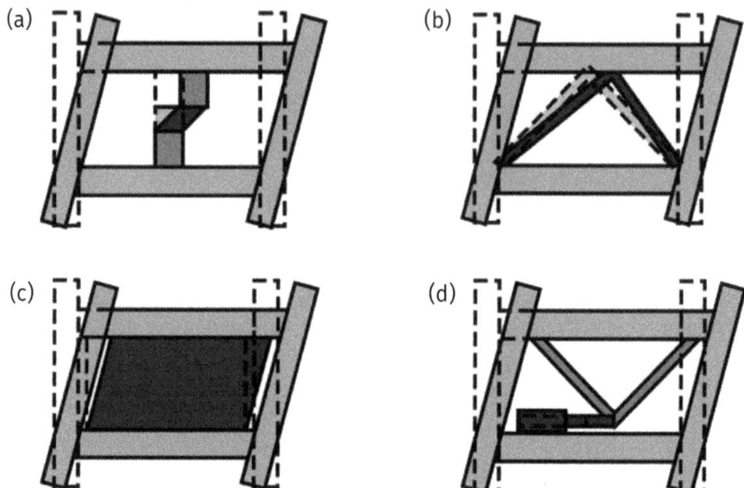

Fig. 2.16 Typical configurations of energy dissipation devices within moment frames: (a) Column, (b) brace, (c) wall and (d) shear link [32]

frames at core areas or perimeter zones of the buildings in order to avoid interference with the building's functions.

The efficiency of energy-dissipation devices is affected by the stiffness ratio of the dampers to the moment frames. If the moment frame is substantially stiff, then the damper will not be effective until a large drift has been reached. Reinforced concrete moment frames typically have high stiffness and, therefore, dampers and their connections must also have high stiffness so that they are engaged and start dissipating energy at small storey drifts.

In most international seismic codes, energy dissipation devices are basically divided into two types, namely, displacement-dependent devices and velocity-dependent devices. Other resources classify the dampers as metallic (hysteretic) dampers, viscous dampers, and viscoelastic dampers. A comparison table showing the advantages/disadvantages of several response spectrum damper types is given in sub-section 2.4.

The most commonly used hysteretic dampers are metallic dampers: Buckling-Restrained Braces (BRB) and friction dampers, whose main characteristic properties are described below.

As for metallic dampers, compared to axial buckling, shear buckling is relatively stable in the post-yield/plastic phase. Exploiting this characteristic, steel shear panels are used as elastoplastic dampers. An example layout of panel damper, an application photo, and typical hysteretic behaviour are shown in Figure 2.17, where shear stress (τ) normalised to the yield shear stress (τ_y) is presented. The panels are composed of low-yield strength steel plates with welded restraining ribs/stiffeners. Configurations are possible with wall-type and column-type panels.

The Buckling-Restrained Brace (BRB) is a brace-type hysteretic damper with elastoplastic behaviour. It consists of an axially yielding steel core and an axially-decoupled restraining mechanism, which suppresses overall buckling. As shown in Figure 2.18(a),

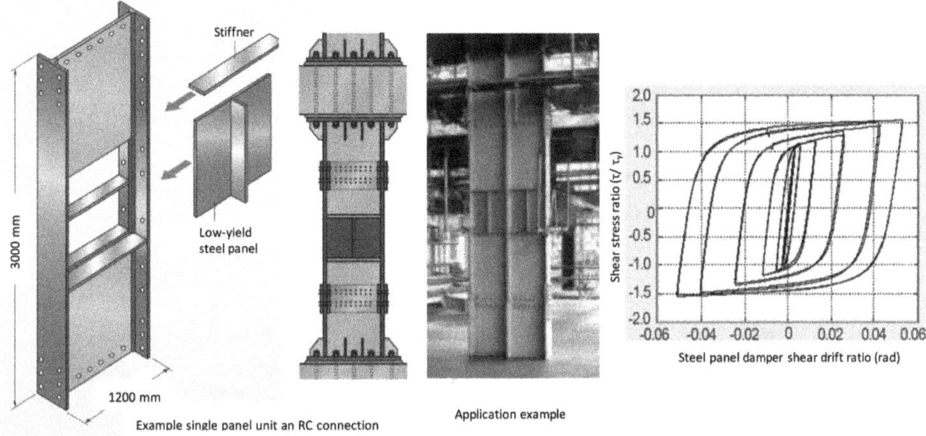

Fig. 2.17 Column-type steel shear panel damper (Photo: JFE) [33]

the buckling restraining mechanism is often composed of a hollow steel section with in-filled mortar and a debonding layer surrounding the yielding core. Other types of restraint shapes and material combinations are shown in Figure 2.18(b). A typical BRB application in chevron configuration is shown in Figure 2.18(c). Under compressive demands and prior to yielding, the core will start to buckle with progressively higher mode shapes developing due to the presence of the restraining mechanism. After the core yields, it exhibits very stable and symmetric hysteresis between compression and tension. These energy dissipation characteristics are excellent and compare favourably against other fully ductile systems, as shown in Figure 2.18(d). For this reason, BRBs can be employed as hysteretic dampers. BRBs are widely used not only in steel structures but also in reinforced concrete structures and often as a seismic retrofit measure.

A friction damper consists of brake pads in contact with a stainless-steel surface. Spring-loaded bolts maintain constant pressure on their interface, as shown in Figure 2.19(a) and (b). A typical friction damper hysteretic behaviour is shown in Figure 2.19(c), where the characteristic vertical cuts in maximum displacement, representing the change in the friction load direction, are visible. Although it is a displacement-dependent energy dissipation device (together with shear panel dampers and BRB's), its initial stiffness is relatively high, and low-cycle fatigue does not occur. This type of damper is suitable for retrofitting RC structures due to the fact that energy is dissipated from low levels of the storey drift [34].

The next type is viscous dampers, which can be produced in brace and wall types. Figures 2.20(a) and (c) show an energy-dissipation device called a viscous wall-type damper component in cross-section and an actual application. Typical cyclic behaviour from a prototype test of a viscous wall damper is shown in Figure 2.20(b). This device comprises a thin-walled void that is filled with viscous fluid. The fluid-filled void is connected to a beam at floor level. An internal steel plate sits within the fluid-filled void and is connected to a beam on the floor above. As already mentioned, there are also braced viscous dampers. Two characteristic examples of a shear link and a toggle-brace configuration are presented in Figure 2.21(a) and (b), respectively.

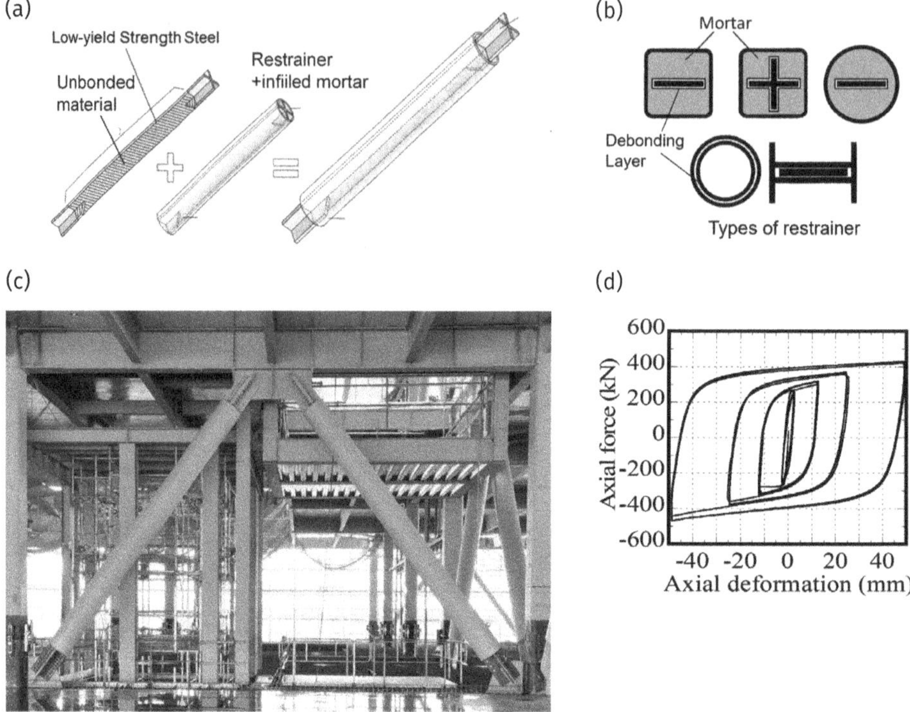

Fig. 2.18　Buckling-Resistant Braces (BRB): (a) Concept of BRB, (b) types of restraining mechanism, (c) typical BRB application (Photo: T. Takeuchi) and (d) hysteresis of a well-designed BRB [35]

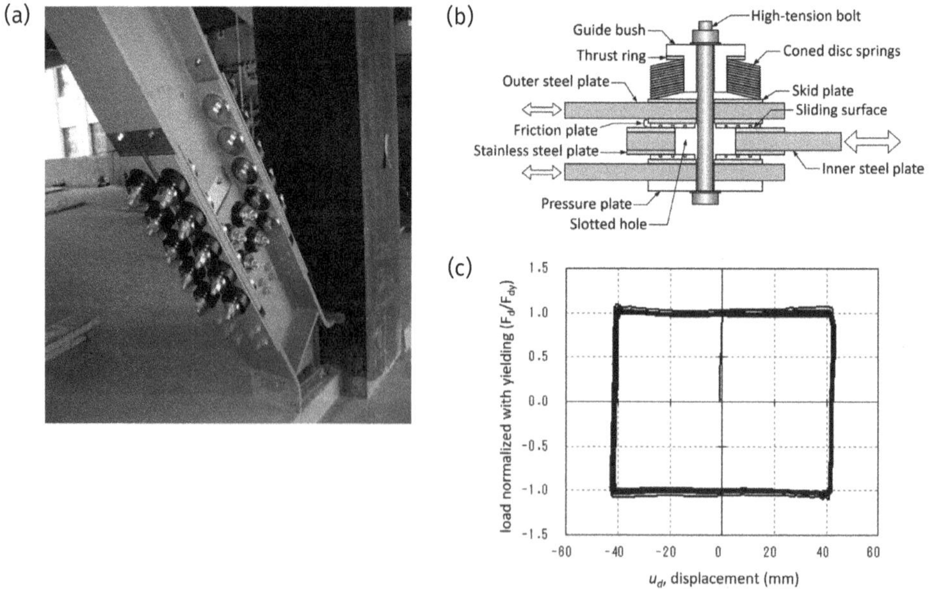

Fig. 2.19　(a) Friction damper, (b) connection detail and (c) typical hysteresis [34]

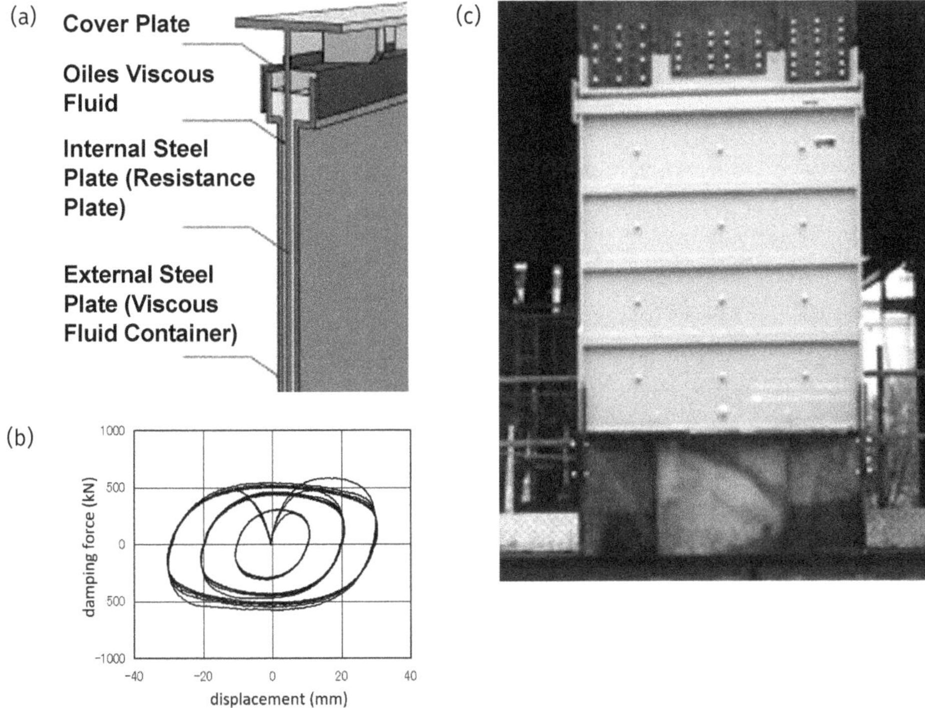

Fig. 2.20 (a) Technical components, (b) typical hysteresis [36] and (c) characteristic application of a viscous wall (Photo: Oiles Corporation Japan)

Fig. 2.21 Viscous damper: (a) shear link configuration and (b) toggle-brace configuration (Photo: F. Sutcu)

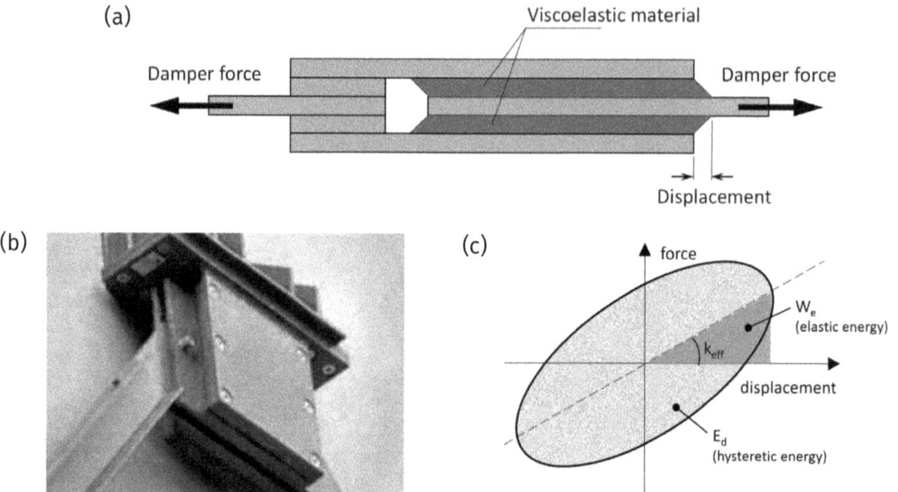

Figure 2.22   (a), (b) Viscoelastic damper (Photo: FIP Industriale [40]) and (c) typical hysteresis

Viscoelastic dampers consist of solid elastomeric pads (viscoelastic material) bonded to steel plates, where brace type and wall type configuration is available. A cross-section drawing and an application photo are shown in Figure 2.22(a) and (b), respectively. The typical inclined elliptical hysteretic behaviour model is shown in Figure 2.22(c). As one end of the damper displaces relative to the other (Figure 2.22(a)), the viscoelastic material undergoes shearing action. This results in the development of heat, which is dissipated to the environment. By natural material properties, viscoelastic dampers exhibit both elasticity and viscosity, which means they are displacement- and velocity-dependent. It should be noted that the behaviour of viscoelastic dampers is also highly dependent on temperature. With increasing research in the field, the characteristics and the performance of the response control devices are continuously improved, and at the same time, more reliable models are developed for the analysis and the design of these systems. A series of research papers describe the state of the art on the design of these systems [37]–[39].

2.4 Summary / Comparison of the Presented Seismic Isolation and Response Control Systems

In the current section, the main advantages and disadvantages of the seismic isolation and response control devices presented in sections 2.2 and 2.3 are summarised. Table 2-1 provides the comparison of the advantages and disadvantages of each type of seismic isolation and response control device, along with a short summary of their objective of use. Furthermore, in Table 2-2, indicative hysteretic force-deformation behaviours of the most common seismic isolation and response control devices are presented, which may assist the selection process for the most suitable device.

Table 2.1 Comparison of various seismic isolation and response control devices

	Type	Objective of Use	Advantage / strong point	Disadvantage / challenge
Seismic isolation devices	Natural Rubber Bearings (NRB)	Seismic isolation of buildings or bridges, both for new design and retrofit	• Reliable and stable material • Maintenance-free • Limited tension capacity • Reserve displacement capacity	• Additional devices with damping capability • Demanding design process • Not suitable for light structures
	Lead-plug Rubber Bearings (LRB)			• Demanding design process • Heavy metal core • Not suitable for light structures
	High Damping Rubber Bearings (HDRB)			• Demanding design process • Not suitable for light structures • Special care should be paid to axial tension forces
	Friction Pendulum Systems (FPS)		• Long period even for light structures • Period independent of the structure's mass • Translational modes of vibration • Easy design.	• Vulnerable to environmental effects • Uplift risk (no tensile capacity) • Dust protection required
	Spring type isolators		• Vertical isolation	• Not suitable for heavier structures • Not suitable to achieve long periods
	Cross linear guide bearings	Supplemental device for improving the seismic response of a seismically isolated device	• Increased structural period with minimal friction	• Maintenance required
	Slicers with rubber		• Increased structural period with limited stiffness and rotation capacity	• Dust protection required • Uplift risk (no tensile capacity)
Response control devices	Elastoplastic dampers such as BRBs or steel panel dampers	Improving the stiffness and damping properties of new and existing buildings or bridges	• Force-limited • Easy to construct • Relatively inexpensive • Adds both "damping" and stiffness	• Highly nonlinear behaviour • Adds stiffness to the system • Undesirable residual deformations possible
	Viscoelastic dampers		• High reliability • May be able to use linear analysis • Somewhat lower cost	• Strong temperature dependence • Lower force and displacement capacity • Not force-limited • Necessity for nonlinear analysis in most practical cases
	Friction dampers	Improving the damping properties of new and existing buildings or bridges	• Force-limited • Easy to construct • Relatively inexpensive	• Highly nonlinear behaviour • Adds large initial stiffness to system • Undesirable residual deformations possible
	Viscous dampers		• High reliability • High force and displacement capacity • No additional stiffness • Simpler design	• Higher cost • Not force limited (particularly when exponent=1) • Necessity for non-linear analysis
	Tuned Mass Dampers	Improving the damping properties of new and existing high-rise structures or bridges	• Reduce vibrations in service condition • Effective for wind loads and long-period seismic effects • Minimum interruption in case of retrofit	• Not effective for strong motion • Not suitable for short or mid-rise buildings

Table 2.2 Comparison of various seismic isolation and response control devices in terms of conceptual force-deformation behaviour

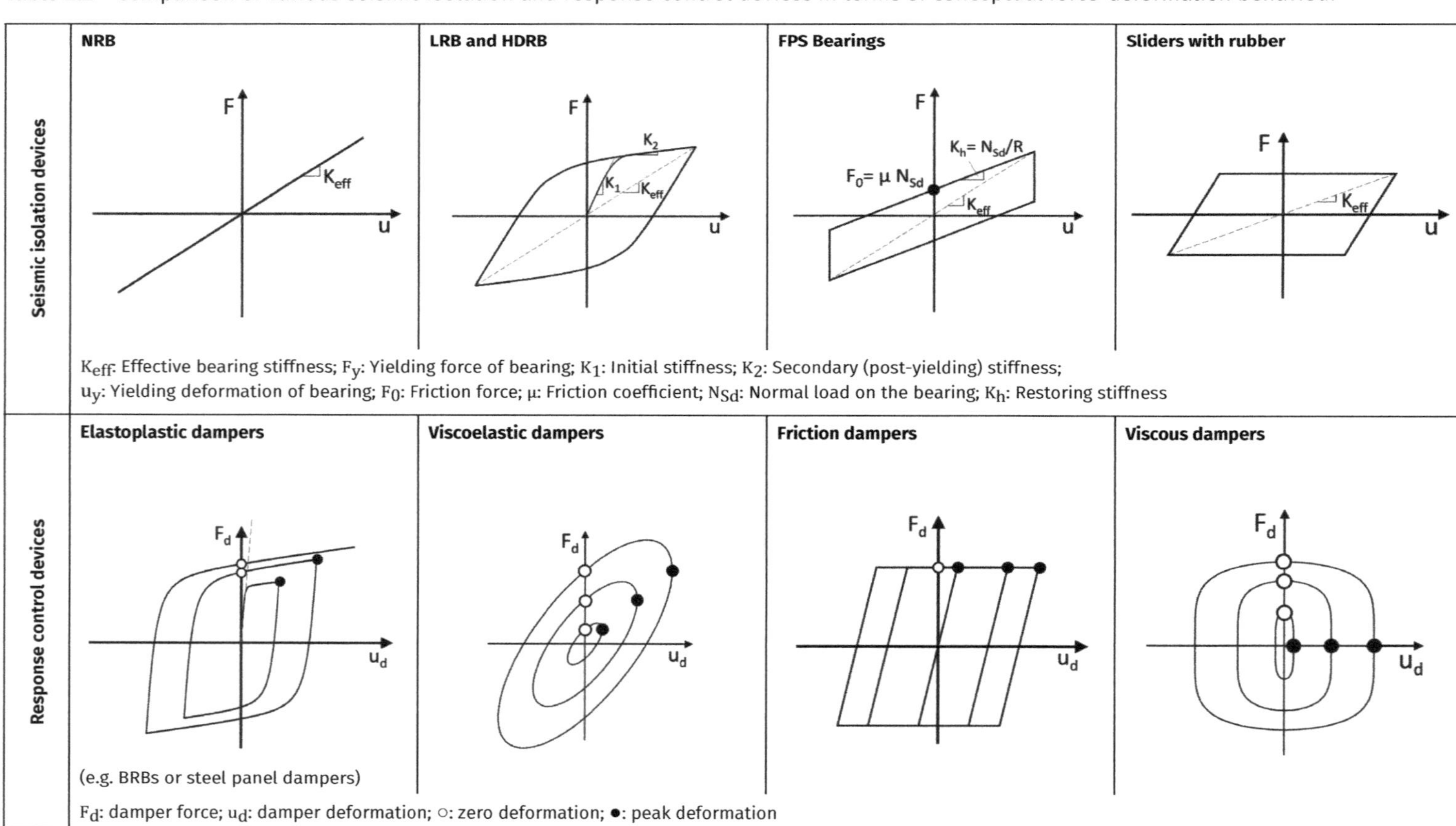

Chapter 3

Design of New Buildings with Seismic Devices

In this chapter, the design of seismic devices for new structures is discussed. The main international code provisions are summarised, followed by a description of the basic design philosophy and representative case studies.

3.1 Design of New Buildings with Seismic Isolation

In this section, the basic descriptions of the main design codes and recommendations worldwide concerning seismic isolation systems (i.e. Eurocode, ASCE, Japanese, Mexican, and Turkish building codes) are presented.

EN 15129 [41] covers the design of seismic isolation devices that are assembled in structures, with the aim of modifying their response to the seismic action. It specifies functional requirements and general design rules of the devices for seismic and non-seismic design situations, material characteristics, manufacturing and testing requirements, as well as assessment and verification of constancy of performance, installation and maintenance requirements. This European Standard covers the most important types of devices and their combinations. In EN15129 [41], seismic isolation devices are divided into three categories:

1) Elastomeric Isolators (Chapter 8.2 of [41]): These are divided into High Damping Rubber Bearings (HDRB) and Low Damping Rubber Bearings (LDRB). LDRB can include Lead-plug or Polymer plug (LRB or PRB) for achieving the desired level of damping.

2) Curved Surface Sliders (Chapter 8.3 of [41]): These are friction pendulum system (FPS) bearings that dissipate energy by friction and provide a restoring force depending on displacement for re-centring.

3) Flat Surface Sliders (Chapter 8.4 of [41]): These sliders should be used with other devices that provide re-centring.

According to EN 15129 [41], it is not recommended to use displacement limiting/stopping rings in sliding devices. That is why triple pendulum bearings are not compatible with this code.

Eurocode-8, Chapter 10 "Base Isolation", covers the design of seismically isolated structures in which the isolation system, located below the main mass of the structure, aims at reducing the seismic response of the lateral-force resisting system [42]. In this code, there is also a brief chapter describing the combination of seismic isolation bearings with damping devices.

In US code, ASCE 7-16 [43], isolators are not classified in the main code, but they are only mentioned in the Commentary section C17. Even in this section, there is no clear classification considering the materials of isolators. Isolators are classified according to their hysteretic behaviour.

In Japan, seismic isolation devices, including laminated rubber bearings and dampers, are defined in Japanese Standards law notification No.2009 (2010), which requires certification by the Ministry of Land Infrastructure Transportation and Tourism (MLIT) before applying them to actual buildings [44]. Minister's certifications are issued after examining test results and specifications by the expert committees.

In New Zealand, there is currently no code in effect that addresses seismic isolation systems. There is a draft for trial use [45], so usually, ASCE 7 [43] is adopted as a substitute.

In Mexico, the Design Manual for Civil Structures: Seismic Design (CFE, 2015) includes design recommendations for elastomeric bearings, LDRBs and sliding isolation systems [46]. Even though this Manual is not a regulation code, it is widely used to design structures in the country.

In the Turkish Building Seismic Code 2018 [47], the seismic isolation devices are mainly categorised into two types, namely, elastomeric isolation bearing units and friction pendulum bearing units. Then, the elastomeric bearings are classified into elastomeric bearings with lead-plug and high-damping elastomeric bearings.

International codes from Europe, the US, Japan and Turkey are compared in terms of essential parameters and rules of seismic isolation design in Table 3-1.

Table 3.1 Comparison of essential code properties for seismic isolation (Comment list of Abbreviations)

Code		EUROCODE [42]	US CODE [43]	JAPANESE CODE [44]	TURKISH CODE [47]
Design Method		ELFM/RSA /NLTHA	ELFM/RSA /NLTHA	ELFM/NLTHA	ELFM/RSA /NLTHA
Return Period		475	MCE (2475 or ε=1)	50/500	475/2475
Design Spectra	For Isolator	DBE	MCE	DBE	DD2: 10 % in 50 yrs
	For Building	DBE	DBE=2/3 MCE	DBE	DD1: 2 % in 50 yrs
Importance Factor		Yes ($\leq$1.4)	No	Yes	No
Vertical Component		Considered	Considered	Considered	Considered
Ageing/dispersion		From tests	From tests	1.2	From tests
Safety Factor on Isolation Capacity		1.2 (Reliability Factor)*	Implicit in the MCE design level	Elastomeric: 1.25 sliding/friction=1.11	Implicit in the DD1 level

Torsion factor in ELFM	Calculated	Calculated	1.1	Calculated
Eccentricity limit	2.5 %	-	3 %	5 %
BLD requirements	Low ductility (R≤1.5)	Low ductility (R≤2)	Elastic	Limited ductility (R≤1.5)
Modelling	2D for ELFM, 3D for NLTHA	2D for ELFM, 3D for others	2D	2D for ELFM, 3D for others

*Safety factor in Eurocode is applied because there is only one level of seismic action (return period 475). However, this may lead to more conservative results.

The limitations of the Equivalent Lateral Force Method of analysis to be used for seismically isolated buildings from seismic codes of Europe, US, Japan and Turkey are summarised in Table 3-2:

Table 3.2 Comparison of limitations for Equivalent Lateral Force Method

Code	EUROCODE [42]	US CODE [43]	JAPANESE CODE [44]	TURKISH CODE [47]
Limitation on Site Seismicity	–	$S_1 \leq 0.6$ g	–	-
Limitation on Site Class	–	A, B, C, D	12	ZA, ZB, ZC, ZD
Max. Plan Dimension (m)	50	–	–	–
Max. Building Height (m)	-	20	60	20
Max. Number of Stories	-	4	–	4
Isolation Units Location	Above substructure	–	Columns' Base	-
Eccentricity Limitations	3 %	–	3 %	5 %
Kv/Ke	≥150	–	–	–
Period Range of Isolated Structure	$3T_{fixed}$–3 s	$3T_{fixed}$–3 s	$T_{isol} \geq 2.5$ s	≤ 4.0 s
Max. Vertical Period (T_v)	$T_V < 0.1$ s	–	–	$T_V < 0.1$ s

3.2 Basics of Seismic Isolation Design

The effect of seismic isolation on the dynamic behaviour of a structure can easily be explained by referring to the acceleration and displacement response spectra. The addition of a seismic isolation layer increases the vibration period of a structure. On the acceleration spectrum, an increased period results in a decreased acceleration response. Meanwhile, the additional damping provided by the isolation layer further reduces the acceleration response (Figure 3.1(a)). As for the displacement response, an increased period naturally brings an increased displacement response. Similar to the acceleration response, the displacement response is also reduced by the additional damping of the isolation layer (Figure 3.1(b)). It should be noted that the spectral displacement re-

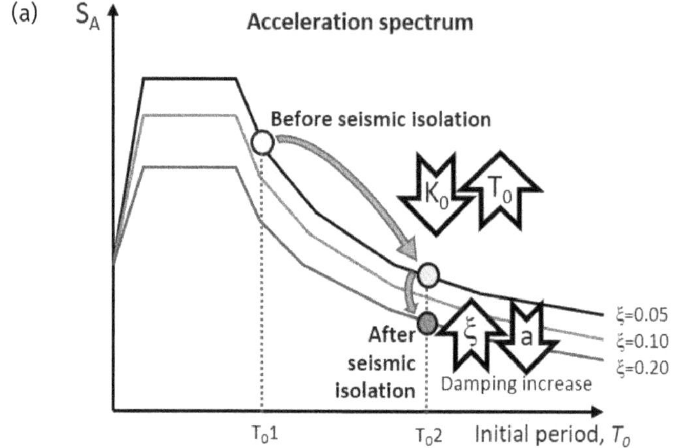

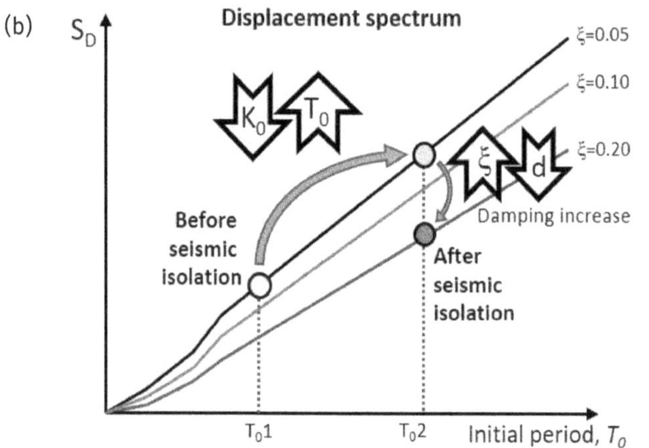

Fig. 3.1 Effect of seismic isolation on the dynamic behaviour of a structure: (a) Acceleration and (b) displacement spectrum

sponse corresponds to the displacement of the isolation layer and not the structural displacement. Seismic isolation devices are manufactured and designed to undergo large displacements.

In the current section, the basic design approach of a seismic isolation system will be explained on the basis of an example, implementing NRB and two different types of dampers. When NRBs are combined with dampers, the hysteretic behaviours should be combined, where the resultant hysteretic model can represent almost any type of isolation system such as LRB, HDRB or FPB. In design, laminated natural rubber bearings are normally modelled as simple horizontal springs. They exhibit a stable linear shear force-horizontal displacement relationship until they start hardening at a displacement approximately equal to 60 % of their diameter, as shown in Figure 3.2.

Viscous dampers are often used with rubber bearings in the seismic isolation layer (Figure 3.3(a)). Viscous dampers generally exhibit velocity-dependent oval hysteresis, as

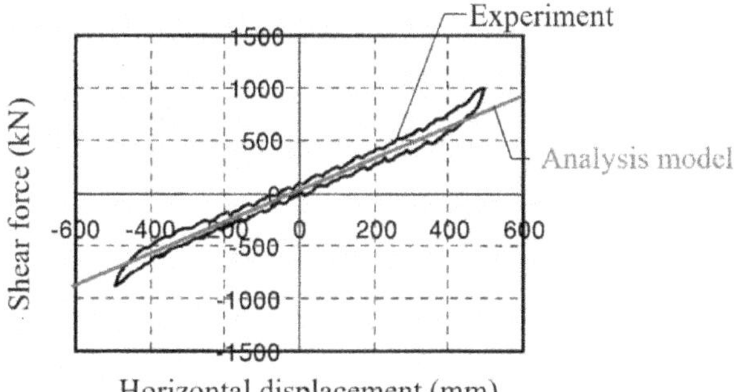

Fig. 3.2 Q-δ relationship of laminated natural rubber bearing with 60 cm diameter [27]

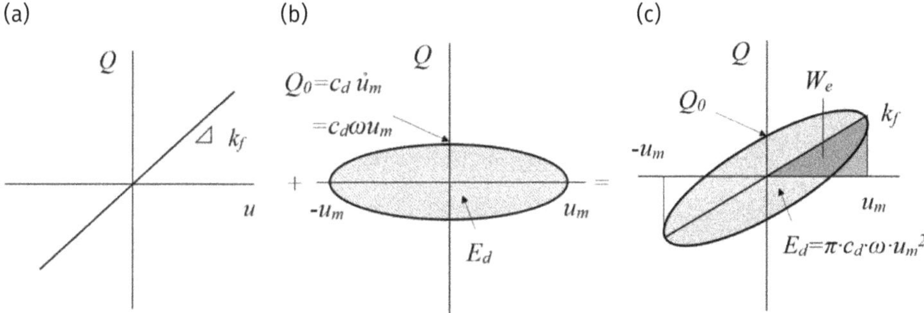

Fig. 3.3 Equivalent damping of viscous system: (a) NRB, (b) viscous damper and (c) viscous system

shown in Figure 3.3(b). When simplified to a linear viscous damping model, their equivalent damping ratio of seismic isolation layer, ξ_v, with liner stiffness k_f (rubber bearings) can be expressed as follows (Figure 3.3(c)) and, in this case, Eq. (3.1) can be converted to Eq. (3.2) by substituting the main parameters.

$$\xi_V = \frac{E_d}{4\pi W_e} \tag{3.1}$$

$$\xi_V = \frac{\pi c_d \omega u_m^2}{2\pi k_f u_m^2} = \frac{c_d \omega}{2k_f} \tag{3.2}$$

Where:

   ξ_v : Equivalent damping ratio of the seismic isolation layer
   E_d : Energy dissipated by the dampers in the seismic isolation system
   W_e : Elastic energy
   c_d : Viscous damping coefficient
   ω : Circular frequency
   u_m : Maximum displacement of the seismic isolation system
   k_f : linear stiffness of rubber bearings

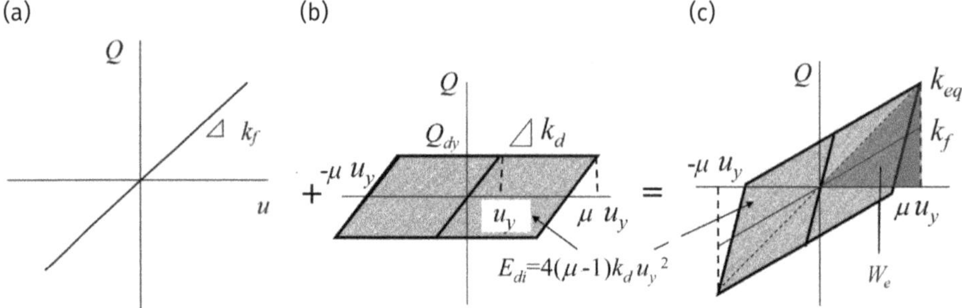

Fig. 3.4 Equivalent damping of elastoplastic system: (a) NRB, (b) elastoplastic damper and (c) elastoplastic system

Similar to the case with viscous dampers shown above, when an elastoplastic type of damper is used with natural rubber bearings, the equivalent damping ratio of this system can also be evaluated from the cyclic dissipated energy (area of hysteresis) E_d as shown in Figure 3.4 and Eq.(3.3).

$$\xi_d = \frac{E_d}{4\pi W_e} = \frac{4(\mu - 1)k_d u_y^2}{2\pi k_{eq}(\mu u_y)^2} = \frac{2\,(\mu - 1)\,k_d/k_f}{\pi\mu(\mu + k_d/k_f)} \tag{3.3}$$

Where:
   ξ_δ : Equivalent damping ratio of the seismic isolation layer
   μ : Ductility ratio
   k_d : Initial damper stiffness
   k_{eq} : Equivalent stiffness of the system
   u_y : Yielding displacement of elastoplastic dampers
   k_f : linear stiffness of rubber bearings

The equivalent damping ratio of an elastoplastic system is displacement (or yield ratio) dependent, and the value obtained by the above equations is based on constant amplitude. Therefore, considering a seismic motion inducing random response amplitudes, the value needs to be reduced by a factor of 0.8~0.6.

   In structural design, many codes propose analysis methods based on a response spectrum, with equivalent linearization procedures to account for dampers using an equivalent damping value. Alternatively, the Non-Linear Time-History Analysis (NLTHA) method includes detailed hysteresis behaviour of each device. Along with the essential design parameters such as stiffness, damping or seismicity, other key parameters such as test conditions, biaxial displacement, recentring properties or vertical acceleration are also extensively covered by the relevant design codes.

3.3 Seismic Joints and Flexible Connections for Equipment in Isolated Buildings

In the application of seismic isolation, there are special implementation demands, as the isolation layer should have the ability to move. Particular care is required to ensure the free movement of the seismically isolated structure at the seismic joint [12], [31]. If the seismic isolation systems are unable to move freely, the seismic performance of the base-isolated superstructure will be significantly compromised, and the effect of pounding may be particularly detrimental for seismic-isolated structures. It is, therefore, not permissible for either structures or finishes to restrict the movement at the joint. Usually, at the ground floor level, proprietary bridging plates span the joints to accommodate pedestrian traffic, as shown in Figure 3.5 and Figure 3.6.

In addition, special attention is required to the design of the staircases or elevators that may pass through the level of seismic isolation. Apart from the free movement of the joints, the good performance of these elements, especially staircases, is crucial since these constitute the escape routes after an earthquake. Figure 3.7 presents a seismic joint at a service staircase passing through the seismic isolation level at the Onassis Cultural

(a)

(b)

Fig. 3.5 Proprietary bridging plate covering the seismic joint around the seismically-isolated Onassis Cultural Center, Athens, Greece (a) main and (b) side entrance (Photo: C. Giarlelis)

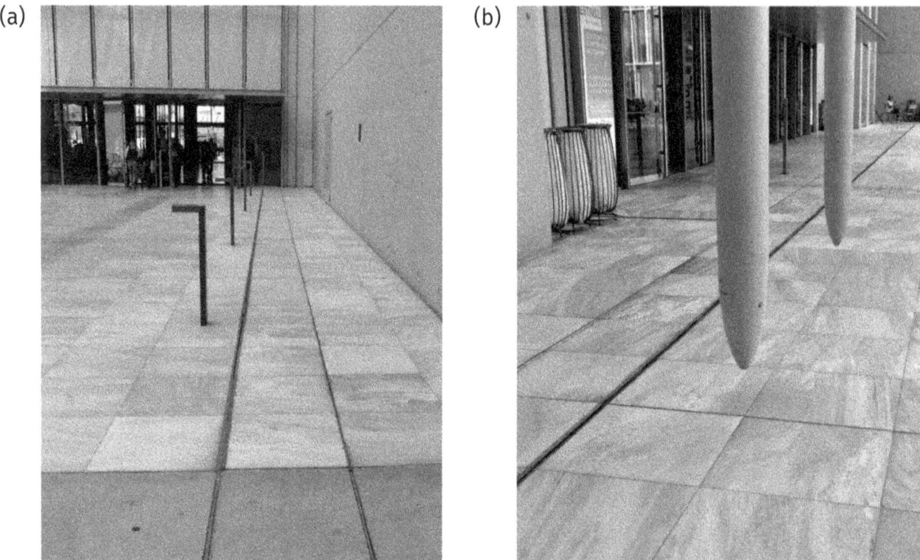

Fig. 3.6   Proprietary bridging plate covering the seismic joint around the Stavros Niarchos Foundation Cultural Center (SNFCC), Athens, Greece (Photo: C. Giarlelis)

Fig. 3.7   Seismic joint at a service staircase passing through the seismic isolation level, Onassis Cultural Center, Athens, Greece (a), (b) bottom and top view of staircase under construction and (c) finished staircase (Photo: C. Giarlelis)

Centre in Athens, Greece, while Figure 3.8 presents another staircase implementation in the Onassis Cultural Centre, where the steel staircase is hanging from the superstructure at the seismic isolation level. Another use of a seismic joint in a staircase is presented in a picture of section 3.4.2.2 (Figure 3.20).

Accordingly, increased care must be taken for lifelines crossing the joints. More specifically, the Mechanical, Electrical and Plumbing (MEP) equipment application in isolated buildings requires special attention according to the movement capacity of these buildings. As a general approach, city mains and the MEP lines in seismically isolated buildings should be connected by specially designed flexible joint members. In addition, the piping/wiring should be implemented to hang from the seismically isolated part of the building (Figure 3.9, Figure 3.10, Figure 3.11, Figure 3.12).

In seismically isolated buildings, expansion joints are essential to cover the clearance between the isolated building and the surrounding moat walls (Figure 3.13). These expansion joints are also used between two building blocks and around elevator shafts. The main function of expansion joints is to allow people to pass over the clearance safely, either between the ground and the building or between two isolated blocks during excitations.

Fig. 3.8 Steel staircase hanging from the superstructure at the seismic isolation level at the Onassis Cultural Center, Athens, Greece (Photo: C. Giarlelis)

Fig. 3.9 Hanging MEP equipment near an isolator at SNFCC, Athens, Greece (Photo: C. Giarlelis)

Fig. 3.10 Hanging piping application in a seismically isolated warehouse project in Japan (Photo: F. Sutcu)

Fig. 3.11 Hanging piping application in a seismically isolated building in Japan (Photo: Dr Y. Shinozaki)

Fig. 3.12 Hanging piping application and extension of pipes and cables between floors in a seismically isolated building project in Japan (Photo: F. Sutcu)

(a)

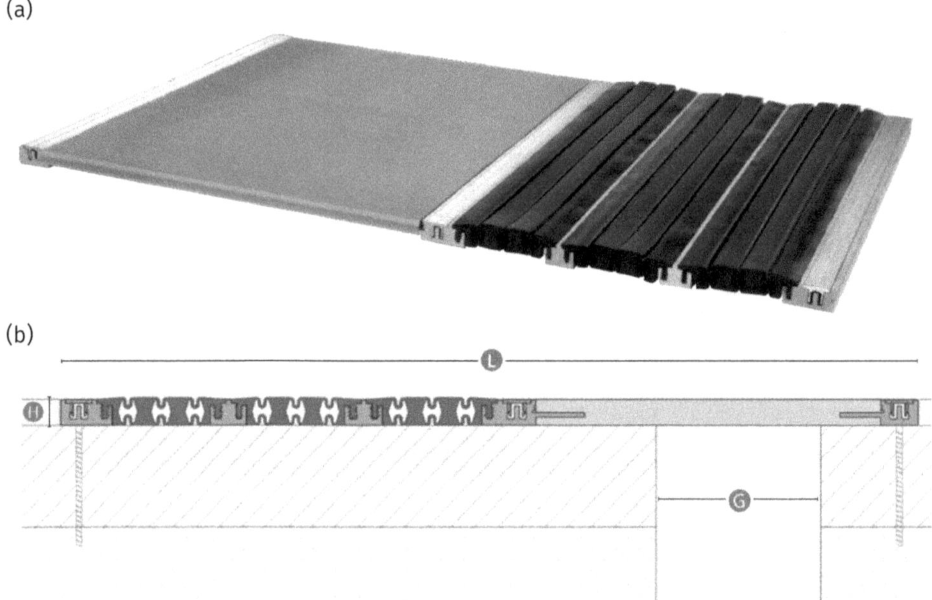

(b)

Fig. 3.13 Expansion joint cover (a) characteristic picture and (b) typical section (Photo: Techno K Giunti) [48]

3.4 Design Examples Using Seismic Isolation

In the current section, four case studies of seismic-isolated buildings in New Zealand, Greece and Mexico are presented where isolation devices of section 2 have been utilised. For each case study, the main objective of the project and the design and performance confirmation of the isolation system are described.

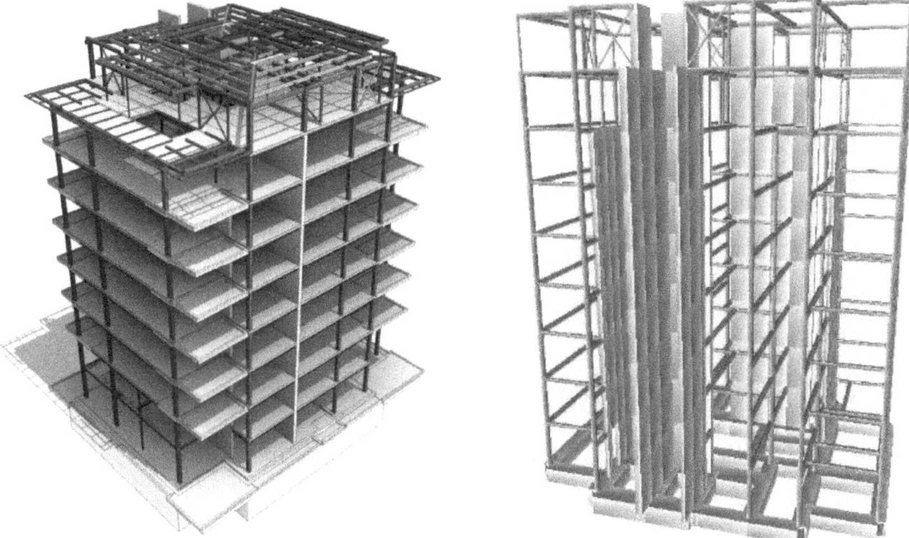

Fig. 3.14   3D model of (a) structural frame and (b) NLTHA model

3.4.1 Nine-Storey Residential Building in Christchurch, New Zealand

The first example is Armagh Apartments, which is a 9-storey residential building located in Christchurch, New Zealand (Figure 3.14). The developer had a clear objective for the building to sustain minimal damage during a 1/500-year earthquake.

3.4.1.1 Objective of the Project

The New Zealand Building Code prioritises life safety over building damage. Therefore, an alternative approach is required, which, in New Zealand, is typically referred to as Low Damage Design (LDD). The objectives of LDD are:

- Structural damage mitigation effectiveness
- Reparability
- Self-centring ability
- Non-structural Damage
- Durability
- Affordability

For a concrete building, structural damage should be limited to minor cracking that can subsequently be repaired by epoxy grout injection. To limit/prevent non-structural damage (i.e. to partitions, building services, façade, etc.), inter-storey drifts should be limited to 0.5 % from a 1/500 year event, which is then referred to as the Damage Control Limit State (DCLS) (Figure 3.15(a)). A base-isolation scheme was adopted for this building in order to meet these main objectives. Lead-rubber bearings were used, which have

(a)

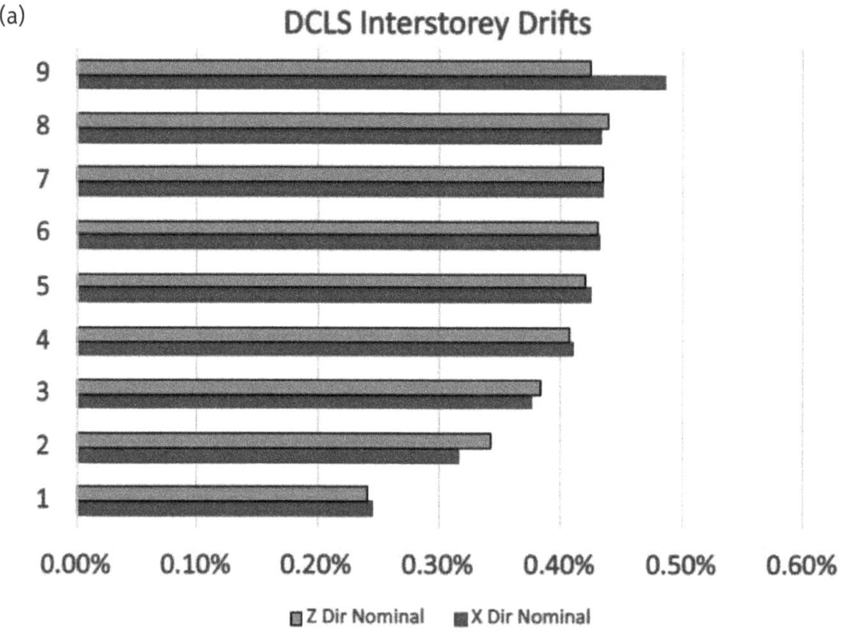

(b)

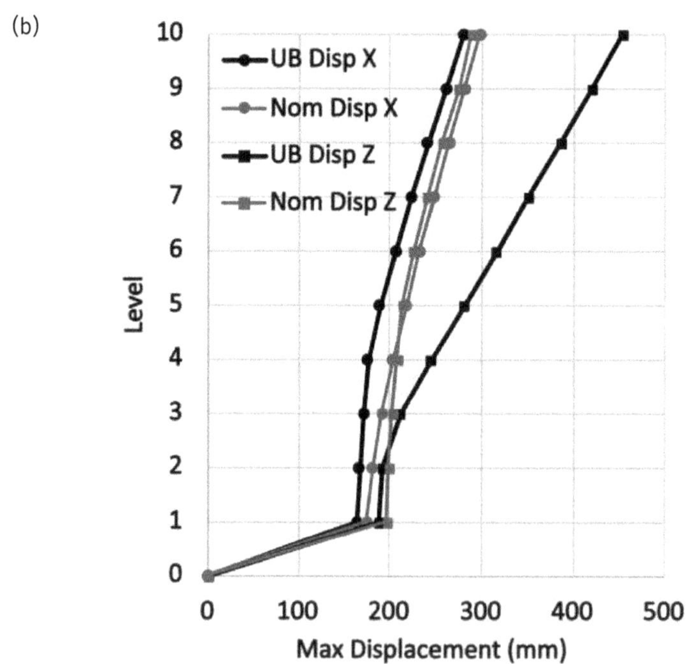

Fig. 3.15 (a) Inter-storey drifts above isolation plane with 1/500 yr event and (b) displacements including isolation plane with 1/2000 yr event ('Nom' denotes analysis with nominal isolation bearing properties and 'LB' denotes analysis with lower bound isolation bearing properties)

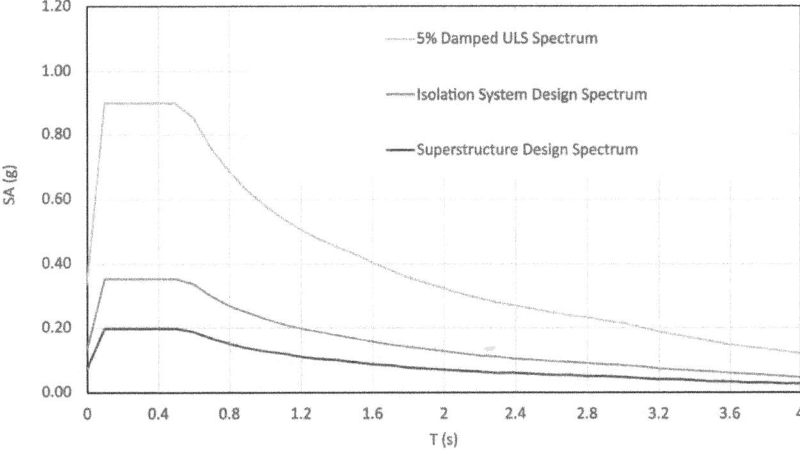

Fig. 3.16 Comparison of design spectra for the isolation system and superstructure with the codified 5 % damped elastic ULS spectrum

excellent self-centring characteristics and in New Zealand can be affordable relative to a conventional structure. Aside from LDD considerations, the building was also checked against collapse limit state criteria with a 1/2000 year earthquake. Displacement of the structure at the isolation plane (Figure 3.15(b)) with this loading was used to determine the size of bearings required and the "rattle space" required around the building to prevent any objects inhibiting the building's movement.

3.4.1.2 Design and Performance Confirmation

An initial isolator layout was developed using a single-degree-of-freedom hand calculation. A preliminary design for the building above the isolation plane was then developed using a modal response spectrum analysis, where the spectrum was adjusted to account for the effects of the isolation bearings (Figure 3.16). Subsequently, a full building non-linear time-history analysis was performed to verify all aspects of the design.

The isolation plane is created between two layers of reinforced concrete structure (Figure 3.17). The upper layer comprises a grillage of beams upon which the building structure is supported. The bottom layer comprises a raft foundation with upstand plinths. The purpose of the beam grillage is to provide a stiff base for the building, through which overturning moments from the eccentricity of the isolator bearings can be resisted. The purpose of the plinths is to create a crawl space through which maintenance staff can access the isolator bearings to carry out routine inspections or replace bearings. The crawl space is enclosed around its side edges by concrete retaining walls and a thin concrete capping slab. The capping slab is formed by a combination of pre-cast paving slabs and an in-situ stitch joint. The "rattle space" is the distance between the edge of the concrete beam grillage and the retaining walls. The capping slab overlaps with the surrounding ground by at least the dimension of the rattle space. This ensures the capping slab provides a cover to the rattle space during a seismic event, which prevents anything from becoming trapped and subsequently damaged by the moving building.

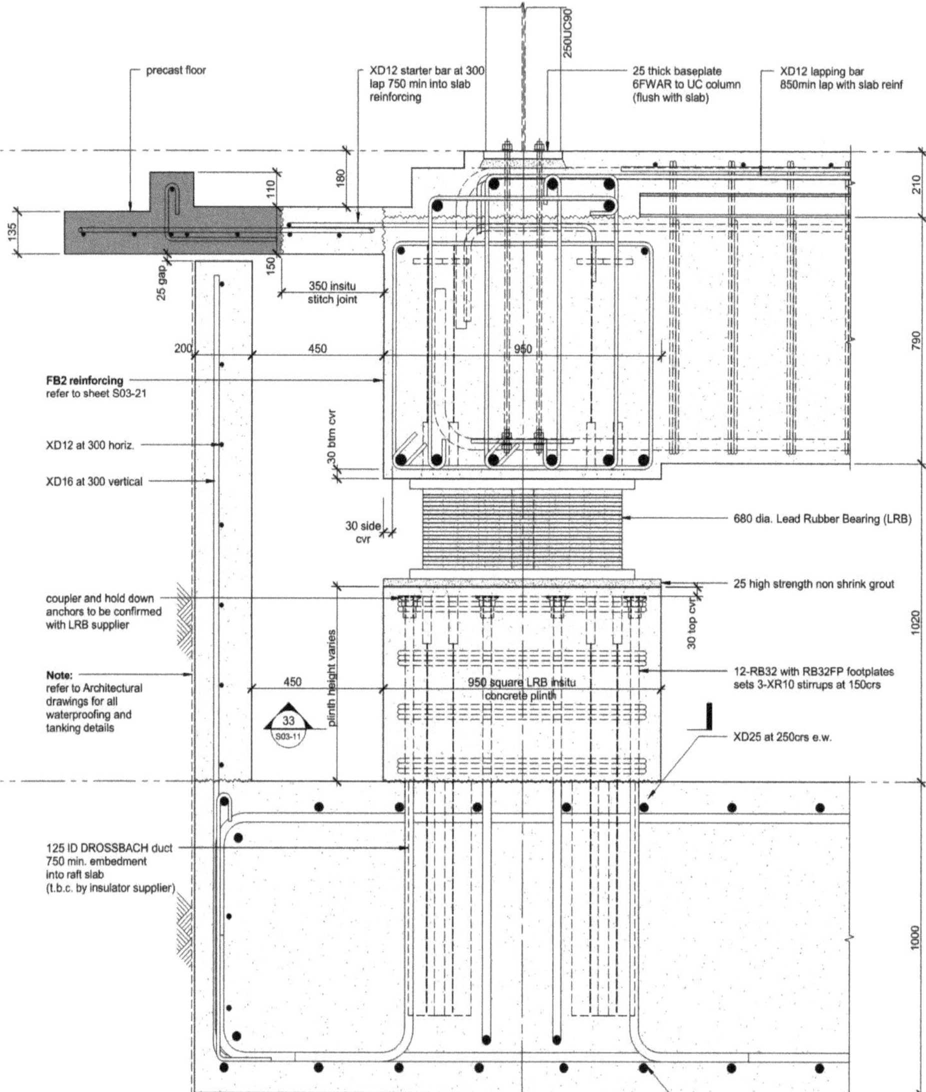

Fig. 3.17 Structural section at isolation plane, showing 450 mm "rattle space" between isolated structure and adjacent retaining walls that form the basement

3.4.2 Stavros Niarchos Foundation Cultural Centre in Athens, Greece

The Stavros Niarchos Foundation Cultural Centre (SNFCC) has been recently constructed in Athens and houses the National Opera and Library of Greece, which started operating in 2017 [12]. The longitudinal section of the cultural centre is presented in Figure 3.18.

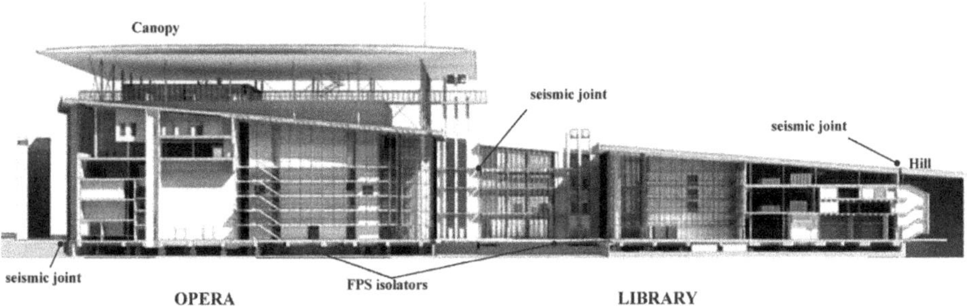

Fig. 3.18 Longitudinal section of the Stavros Niarchos Foundation Cultural Center [12]

3.4.2.1 Objective of the Project

The main structural challenges of the project are the very poor soil conditions (the need to account for soil-structure interaction effects), the high seismic performance expectations and the need to satisfy a demanding architectural concept. Therefore, the incorporation of a seismic isolation system of friction-pendulum type (FPS) was deemed necessary. A total number of 323 bearings for the two buildings were placed at a uniform level beneath the ground floor slab. For the design, Eurocode 8 [42] was used for the first time for a building project in Greece.

3.4.2.2 Design and Performance Confirmation

A site-specific spectrum was deemed necessary in view of the geotechnical conditions. The elastic response spectrum corresponding to a 475-year return period is presented in Figure 3.19. For the design of the superstructure and the isolation system, a damping correction factor of η=0.70 is applied to the elastic design spectrum for the response spectrum (RS) analysis to account for the isolators' energy dissipation. For the isolation system design, the green dashed line of Figure 3.19 is used. It should be mentioned that for the time-history analyses, the damping values corresponding to the friction properties of the isolators are used; these values are lower and correspond accordingly to a lower damping correction factor. Regarding the behaviour factor, a value of q=1.5 is applied for the superstructure design spectrum; this value is the maximum allowed by EC8 for seismic isolated structures and accounts for the overstrength in the design (blue dashed line in Figure 3.19).

For the isolation system design, no behaviour factor is applied (q=1) for checking their stress condition (Figure 3.19). Furthermore, a magnification factor, γ_x=1.5, as ascribed by the Greek National Annex of EC8, is applied to the displacements (thus corresponding to higher seismic demand).

Preliminary investigations of the structures' seismic behaviour showed that without seismic isolation, their fundamental periods lie on the horizontal part of the spectrum, i.e. with corresponding accelerations of 1.12g for elastic design and 0.75g considering a behaviour factor of q=1.5. Using seismic isolation and selecting a period of T=2.59s, results in a design acceleration of 0.093g for the superstructures. Thus, without seismic

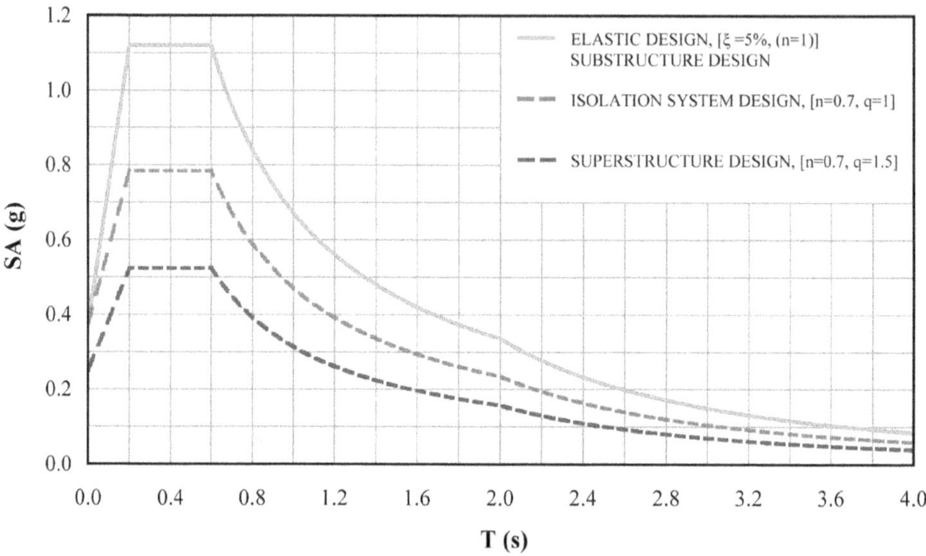

Fig. 3.19   Site-specific horizontal acceleration spectra, $[S{\cdot}a_{gR}{=}0.267, \gamma_I{=}1.4]$

Fig. 3.20   Seismic isolator on its pedestal and seismic joint on a staircase
(Photo: C. Giarlelis)

(a)   (b)

(c)   (d)

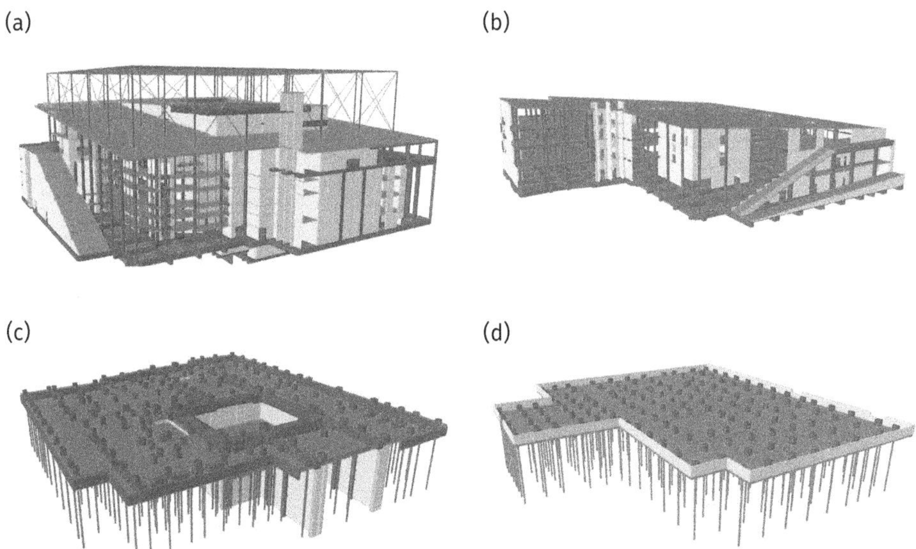

Fig. 3.21 3D model of the (a), (b) superstructures and (c), (d) the foundations of the National Opera and National Library of Greece, respectively (SNFCC) [12]

isolation, their fundamental periods would be seven times higher. The selection of the fundamental period of the structure is related to the lower limit of spectral acceleration imposed by EC8. Following the code, the spectral acceleration should always be higher than $\beta\, a_g$, where β, the lower bound factor for the horizontal design spectrum, is 0.2.

The seismic isolation system is designed for the structure to specifically demonstrate a desired dynamic behaviour. The critical dynamic property is the fundamental period: as the period increases, the spectral acceleration decreases. However, there is a lower limit to spectral acceleration, according to EC8.

For both structures, seismic isolation at the ground floor level separates the super-structure from the substructure, as shown in Figure 3.20, where a typical isolator on its pedestal and a seismic joint on a staircase is demonstrated. For each structure, two separate 3D finite element models (Figure 3.21) were created, one for each of the super-structures and the basement foundations, respectively. This increases the accuracy and simplifies the analysis. Regarding the foundation, a relatively small number of piles were used, approximately one-third of the ones needed if the structures were not isolated, leading to a considerable reduction in the overall cost of the structure. The interaction of the superstructure-foundation-soil was considered using an iterative procedure.

Non-linear analyses, following the provisions of EC8, were used to design the seismic isolation system and check the displacements. In Figure 3.22(a), the displacement history of an isolator near the perimeter of the Opera along x-direction for the Pacoima Dam record is presented. In Figure 3.22(b), the relative displacement history along the x-direction between the roof and the base of the Opera is presented using the Pacoima Dam record of the 1971 San Fernando earthquake (the record was amplitude–and–frequency manipulated to be compatible with the design response spectrum). This reaches

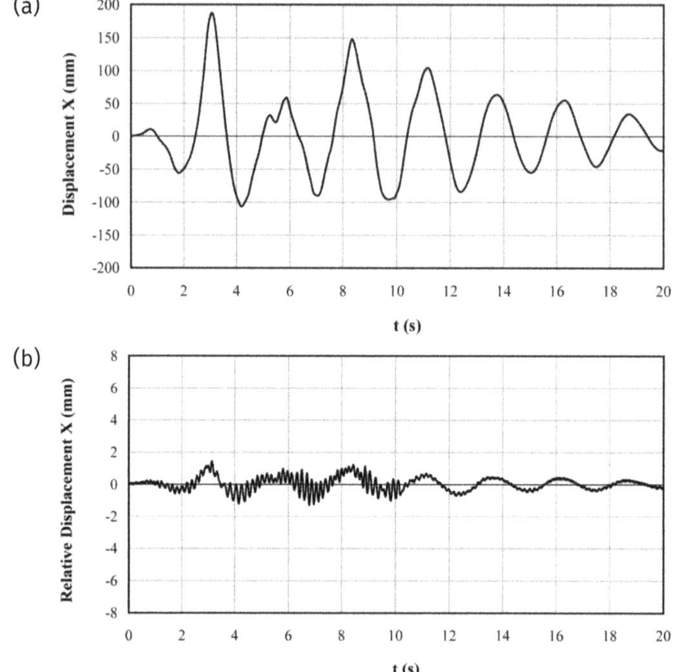

Fig. 3.22 (a) Displacement history along x-direction of an isolator near the perimeter of the
Opera and (b) relative displacement history along x-direction between the roof
and the base of the Opera

a maximum value of 2 mm while the inter-storey drift ratios, γ, are minimal (of the order
of 0.01 %). Therefore, the damage to the structural and non-structural members during
a seismic event is expected to be negligible.

3.4.3 Infiernillo II Bridge, Balsas River, Mexico

The Infiernillo II bridge is one of the first isolated bridges in Mexico. It has a total length
of 525 m with five simply supported 105 m long spans. The bridge width is 12 m, and
the superstructure is supported on steel girders and Camel Back steel trusses. The deck
is composed of a 0.18 m thick reinforced concrete slab. Non-prismatic abutments and
hollow box shape piers compose the substructure (Figure 3.23).

3.4.3.1 Objective of the Project

The close location of the bridge to the subduction zone in the Pacific Coast of Mexico
suggested the use of an isolation system. The bridge is supported on a sliding multi-rota-
tional isolation system (Figure 3.24(a)) composed of disc bearings mounted on urethane
springs. Figure 3.24(b) displays the hysteretic behaviour of the isolation system, which
was calculated numerically.

3.4.3.2 Design Spectrum of the Infiernillo II Bridge

The elastic design spectrum used in the design process of the bridge is illustrated in Figure 3.25 [49]. The total length and location of the bridge classified the structure in Group "A" of the regulation code (importance factor=1.5). Figure 3.25 also shows the reduced spectrum for a behaviour factor $q=2$ and a damping modification factor of 0.52 that was applied for periods larger than 80 % of the isolated fundamental bridge.

3.4.3.3 Design and Performance Confirmation

The bridge was finished in 2003, and since then, several earthquakes have occurred. So far, its seismic response has been adequate without any damage. The seismic vulnerability of the bridge has been assessed in research studies, showing that large return periods of earthquakes can lead to significant damages [50], [51].

Based on a campaign of ambient vibration measurements, the fundamental periods of the bridge in the transverse, longitudinal and vertical directions were 2.5 s, 2.4 s and 0.9 s, respectively. The bridge has a significant irregularity due to the 70-metre high central piers that are much higher than the adjacent piers with heights in the range of 35–54 m.

The analytical studies exhibited that the isolation system favourably contributed to reducing the irregular seismic response of the bridge. As an example, Figure 3.26(a) shows the displacement time-history of piers 2 (54 m height) and 3 (70 m height) subjected to a subduction seismic record (Figure 3.26(b))

(a)

(b)

Fig. 3.23 Infiernillo II bridge crossing the Balsas river in Mexico (Photo: J. Jara)

(a)

(b)

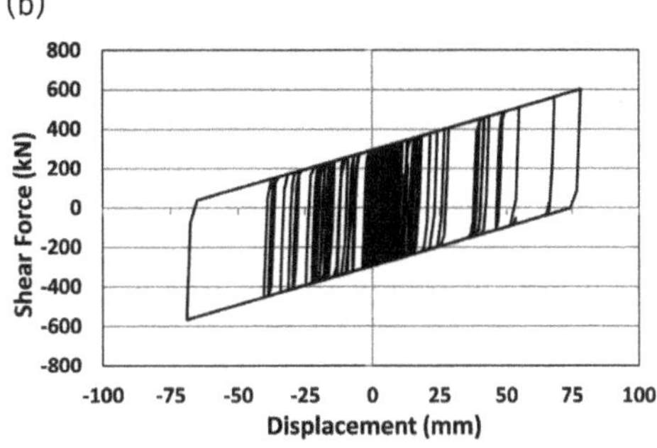

Fig. 3.24 (a) View of an isolator under the deck and (b) hysteretic behaviour of the isolation system (Photo: J. Jara)

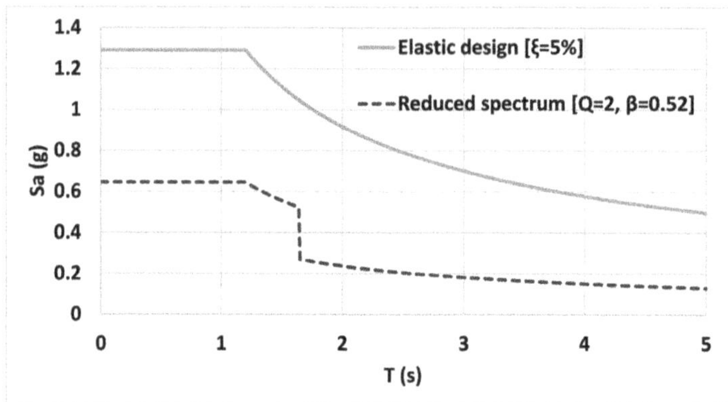

Fig. 3.25   Elastic design spectrum and reduced spectrum used for seismic analysis of the
Infiernillo II bridge

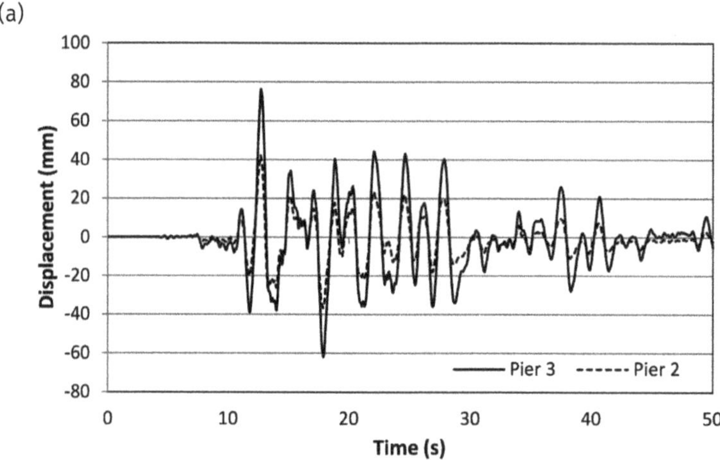

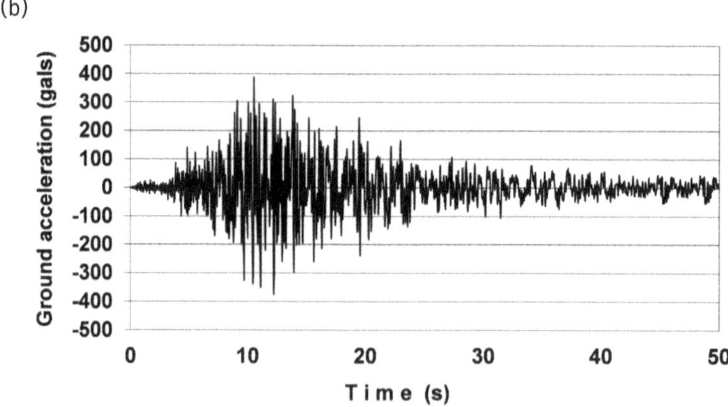

Fig. 3.26   (a) Displacement time-history of Pier 2 and Pier 3 of the Infiernillo II bridge subjected to (b) Manzanillo seismic record (October 9, 1995, Mw=8.0) [51]

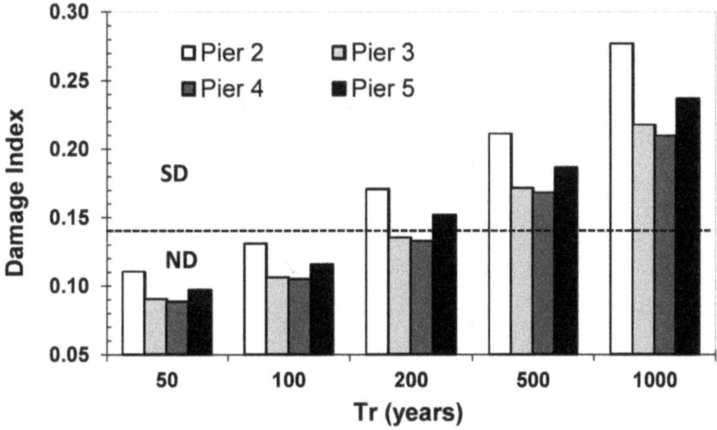

Fig. 3.27 Expected damage indexes for piers of the Infiernillo II bridge (ND = no damage, SD = slight damage) [50]

in the longitudinal direction of the bridge. Even though the different displacement amplitudes in both piers, the elements displayed in-phase movement.

A seismic hazard assessment on-site revealed an expected PGA of 420 gals (4.2 m/s^2) for a return period of 1000 years. Figure 3.27 shows expected damage indexes in each pier of the bridge for different return periods (Tr). To evaluate the expected seismic behaviour of the bridge, damage indices were determined for each pier. One of the most calibrated damage indices used to assess the damage in reinforced concrete elements is the one proposed by Park *et al.* (1984) [44]. The index includes the displacement and hysteretic energy demands additively. A seismic hazard assessment in the site revealed an expected PGA of 420 gals for a return period of 1000 years. Figure 3.27 shows expected damage indices in each pier of the bridge for different return periods (Tr). A horizontal line divides the expected limit states of no damage (ND) and slight damage (SD). The damage limit states were proposed by [52]. Shorter piers (2 and 5) are more vulnerable to suffer damages than taller piers (3 and 4). However, for a return period of 1000 years, none of the piers is in an upper zone of the slight damage index, which proves the efficiency of the isolation system.

Another sliding system used in the Infiernillo II Bridge was manufactured by RJ Watson [53]. This multirotational isolation bearing (Figure 3.28) controls the supplied damping by adjusting the friction in the device. It is composed of steel, urethane, PTFE and stainless steel. The developed system was based on research performed at the Multidisciplinary Centre for Earthquake Engineering Research at the State University of New York at Buffalo.

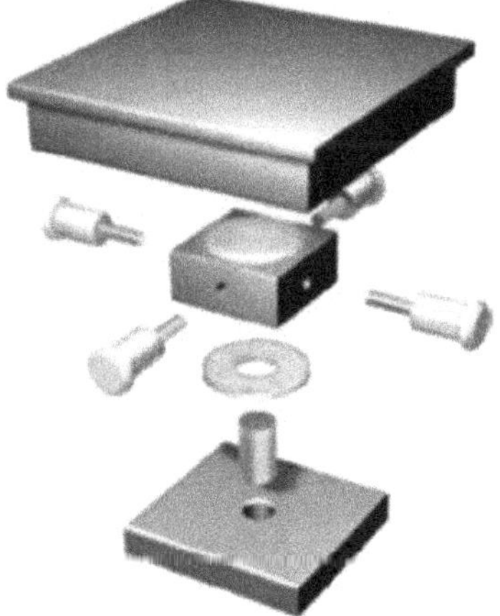

Fig. 3.28 Multirotational isolation system used in the Infiernillo II bridge [54]

3.4.4 Onassis Cultural Center (Stegi), Athens, Greece

The Onassis Cultural Center (Stegi) is a building with landmark architecture. A preliminary structural study indicated that the architectural concept, the need for small size columns on the perimeter to permit a clear view of the egg-shaped interior shell and the presence of marble facades, could not be implemented with the high-performance seismic specifications that were set. In addition, the protection of the contents of the building (i.e. artefacts, recording equipment etc.) from induced seismic accelerations could not be ensured [31].

3.4.4.1 Objective of the Project

The incorporation of a seismic isolation system of FPS type was deemed necessary in order to achieve: operational performance level, protection of the contents of the structure and continuation of its functionality in case of a moderate to strong earthquake. Onassis Stegi is officially the first seismic isolated building in Greece. At the time it was designed (2002–2004), Eurocodes were not available, so, in the absence of a national code for seismic isolation, the International Building Code (IBC) provisions were used.

3.4.4.2 Design and Performance Confirmation

The elastic response spectrum corresponding to a 475-year return period is presented in Figure 3.29. This was used for the design of the superstructure, while, for the isolation system, the Maximum Considered Earthquake, i.e. an earthquake with a return period of 2400 years, was considered following the IBC code requirements. A damping correction factor of $\eta=0.70$ was applied to both spectra for the response spectrum (RS) analysis to account for the isolators' energy dissipation. This factor was applied only for periods larger than 80 % of the fundamental period of the isolated structure, creating the step shown in the response spectra of Figure 3.29. For the time-history analyses, the damping values corresponding to the friction properties of the isolators were used; these values are lower and correspond accordingly to a lower damping correction factor. Regarding the behaviour factor, a value of $q=1.5$ was used for the superstructure design spectrum (blue dashed line in Figure 3.29), while this was taken as $q=1$ for the isolation system design spectrum (green dashed line in Figure 3.29).

The seismic design was performed in three consecutive steps: (1) simple calculations using an equivalent SDOF system, (2) dynamic response spectrum analyses and (3) non-linear time-history analyses using selected earthquake records on a 3-D finite element model, shown in Figure 3.30. The first step is used for the scheme design of the isolation system, i.e. the selection of its dynamic properties. Dynamic response spectrum (R-S) analysis on a 3D finite element model is the next step of the design methodology aiming to: (a) verify the fundamental period, (b) calculate maximum displacements, (c) check for uplift, and (d) do the final design of structural elements. Ultimately, in order to develop additional insight on the performance of the structure but also in order to design the isolation system, nonlinear time-history analyses were performed.

Seismic isolation at the ground floor level separates the superstructure from the basement. The load transmission between the two sections is carried out by 46 bearings placed

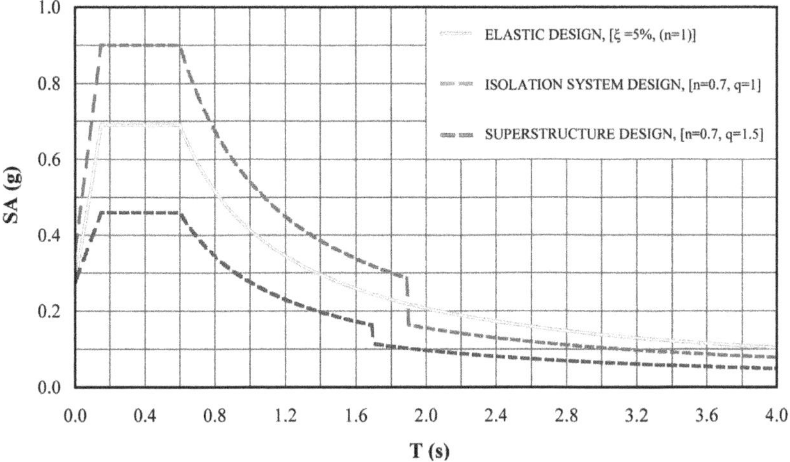

Fig. 3.29 Design horizontal acceleration spectra

(a)   (b)

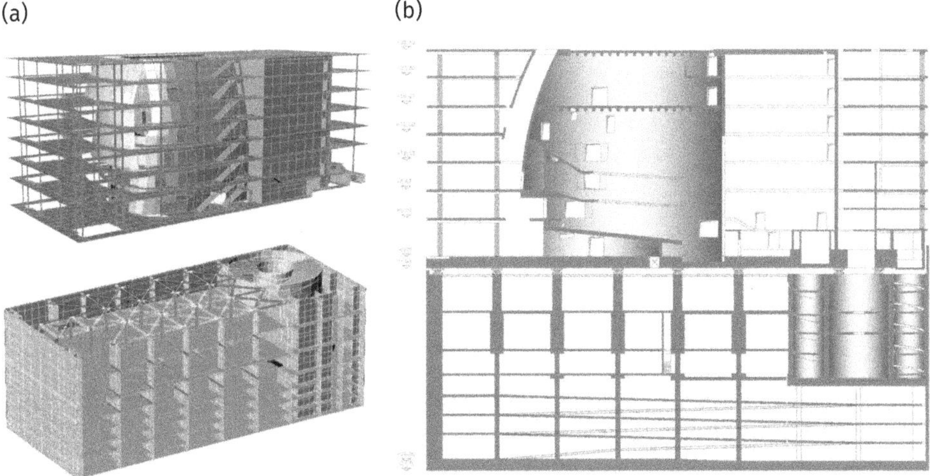

Fig. 3.30 Structural models of the superstructure and the basement: (a) 3D and (b) longitudinal section view.

underneath the ground floor slab (Figure 3.31). The seismic isolation system is designed so that the structure would exhibit a specific desirable dynamic behaviour. The critical dynamic characteristic is the fundamental period; as the period increases, the spectral acceleration decreases, but there is an increase in maximum displacement. Taking into account constraints arising by the size of the seismic joint, a period of T=2.15 s was selected for the seismically isolated structure. This results in a design acceleration for the superstructure of 0.09 g. Had the building been designed without seismic isolation, then, as concluded from preliminary studies, its periods in the horizontal directions would have lain between 0.20–0.34 s (i.e. on the horizontal part of the spectrum), resulting in a spectral acceleration of 0.46 g, which is more than five times bigger. Furthermore, the

Fig. 3.31   Bearings placed on top of the columns of the level -1; the steel truss serves as a diaphragm (Photo: C. Giarlelis)

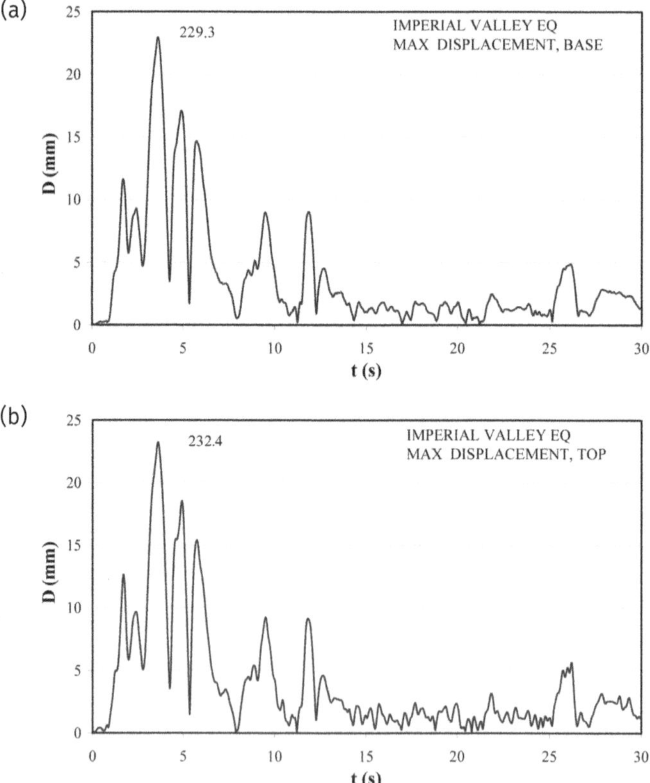

Fig. 3.32   Absolute displacement time-history at (a) the ground and (b) top floor level, respectively

incorporation of the isolation system drastically reduced the inter-storey drift ratios. In Figure 3.32, the absolute displacement time-histories from the modified El Centro record at the ground and top floor level, respectively, are presented. It can be noticed that the difference in the displacement at the ground and top floor levels is minimal. In Table 3-3, the inter-storey drift ratios in directions X and Y are presented. The values obtained for the seismically isolated structure from the response spectrum and the time-history analyses are substantially lower than those for a conventional structure.

Table 3.3 Comparison of inter-storey drift ratio, γ, for different types of analysis

Storey	Height	Conventional structure R-S analysis		Seismic isolated structure R-S analysis		Seismic isolated structure Nonlinear t-h analysis	
	H (m)	$X_{\gamma,max}$ (10^{-3})	$\gamma_{Y,max}$ (10^{-3})	$\gamma_{X,max}$ (10^{-3})	$\gamma_{Y,max}$ (10^{-3})	$\gamma_{X,max}$ (10^{-3})	$\gamma_{Y,max}$ (10^{-3})
7	26.53	0.63	1.13	0.22	0.31	0.20	0.29
6	22.87	0.93	1.44	0.29	0.34	0.17	0.34
5	19.20	0.83	1.85	0.31	0.40	0.19	0.39
4	15.53	0.85	1.85	0.31	0.45	0.19	0.42
3	11.87	0.82	2.11	0.32	0.40	0.15	0.39
2	8.20	1.00	1.76	0.32	0.64	0.17	0.60
1	3.20	0.63	3.01	0.25	0.38	0.17	0.28
0	0.00						

3.5 Design of New Buildings with Response Control

Dampers are supplemental energy dissipating devices used to improve the structural dynamic response of buildings under external loads. For example, dampers are used for response control of buildings subjected to wind or seismic loads [55].

The design of buildings with response control systems is more challenging than the design of seismically isolated buildings. In the case of structures where the damper devices are distributed in the structural plan, dampers can effectively modify structural displacement response and mode shapes. Depending on the damper type, the structural responses can substantially change during seismic motion and, therefore, simple analysis methods, such as equivalent lateral force method or mode combination method, should be applied carefully even within the limits allowed by the relevant codes. Another challenge is that there are various types of dampers, and an effective damper model covering all types of devices is not available.

In the existing international codes, dampers are generally classified as velocity-dependent dampers and displacement dependent dampers. As some dampers' dynamic response is highly dependent on velocity, equivalent static load or mode combination method cannot be accurately used. Consequently, the dynamic time-history analysis method is preferred to understand the actual behaviour and to confirm the overall design of a response-controlled structure.

The American standard, ASCE 7-16 [43], contains a detailed chapter regarding the design of new buildings with dampers. The European standard, EN 15129 [41], covers various seismic devices, including seismic isolators and dampers, where essential design rules are explained, and mechanical requirements for such devices are listed. In EN 15129 [41], although the design section is briefer compared to the ASCE 7-16 [43] code, the requirements of the devices are explained comprehensively. In EC8 [42], a specific section for response-controlled design does not exist. Similarly, the Japanese Seismic Code [44] also does not contain a specific section for response-controlled building design. However, the JSSI Passive Control Manual [32] includes detailed explanations regarding the design of response-controlled structures, including design examples.

3.6 Basics of Response Control System Design

For the design of response control systems, there are two basic steps. In the first step, the structure is converted into an equivalent SDOF system and, for a selected damper type and placement, the total dissipated energy by the dampers is expressed in terms of equivalent modal viscous damping (Eq.(3.3)). In this case, E_d is the dissipated energy in a full cycle for a certain structural mode, and W_e is the elastic energy dissipated at the maximum displacement. The obtained damping is added to the structurally inherent damping, and the design acceleration spectrum is reduced according to this total damping ratio. Depending on the damper type, the period of the structure may change, and this should be addressed in the determination of spectral response evaluation (Figure 3.33 and Figure 3.34).

The building can be designed by a mode combination method implementing the modified spectrum. If the dampers are adding stiffness to the structure, this additional stiffness should be considered in the model. For example, if the dampers in question are displacement dependent, the effective stiffness of the dampers corresponding to the spectral displacement is added to the structural stiffness (Figure 3.33). This design step is similar to the design of conventional buildings using the capacity-demand approach or the preliminary design of seismically isolated structures. Similarly, in the case of velocity-dependent dampers, it should be noted that the spectral response is only affected by equivalent damping, as shown in Figure 3.34.

In the second step, the design is confirmed by non-linear time-history analysis. In this stage, for the confirmation of the dampers, generally, MCE level seismic input (return period of 2475 years) is used. In this stage, the realistic modelling of dampers is crucial for accurate results.

Buckling-restrained Braces (BRBs) are described in this section as a typical response control system. Generally, BRBs are a reasonably well-tested system that can achieve superior performance when the system is carefully designed and detailed by informed engineers and reputable suppliers. However, the unique characteristics of these braces can produce several undesirable failure mechanisms, which are directly influenced by decisions made for the adjacent framing, connections and restrainers. A good system-level design, therefore, requires the engineering consultant to understand the nuances of the brace itself.

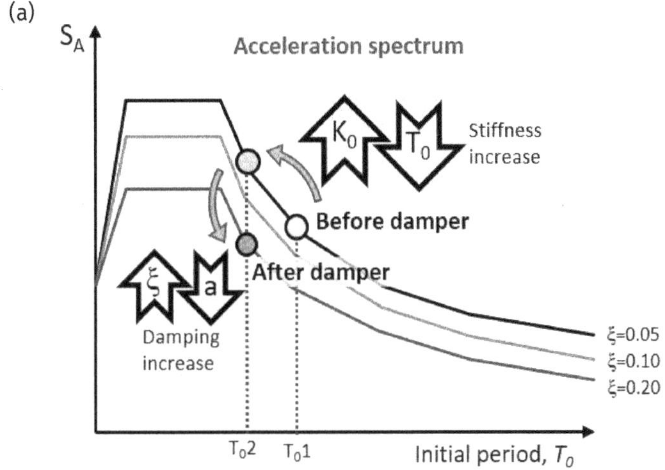

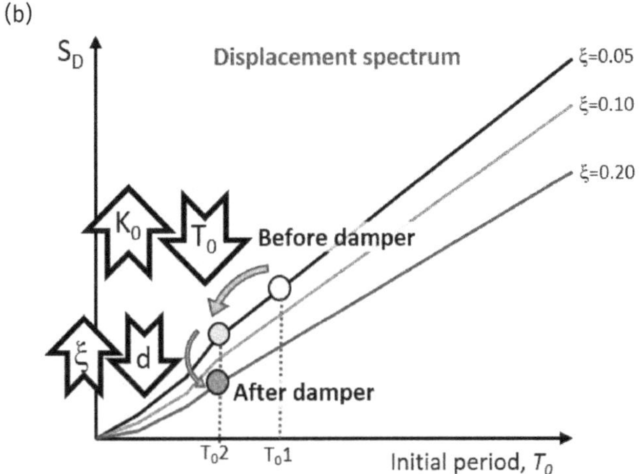

Fig. 3.33 Response control with a displacement dependent device such as BRB: (a) acceleration and (b) displacement spectrum

Detailed studies in recent years have demonstrated particular mechanisms, which can occur at loads and displacements significantly lower than those anticipated by conventional design checks. In general, the BRB must be designed for strength and stability, considering both the local and global behaviour, as shown in Figure 3.35 [56].

To achieve stable hysteresis, the following design conditions should be satisfied:

- *Restrainer* successfully suppresses first-mode flexural buckling of the core;
- *Debonding mechanism* decouples axial demands and allows for Poisson effects of the core;
- *Restrainer wall bulging* due to higher mode buckling is suppressed;
- *Global out-of-plane stability* is ensured, including connections;
- *Low-cycle fatigue capacity* is sufficient for expected demands.

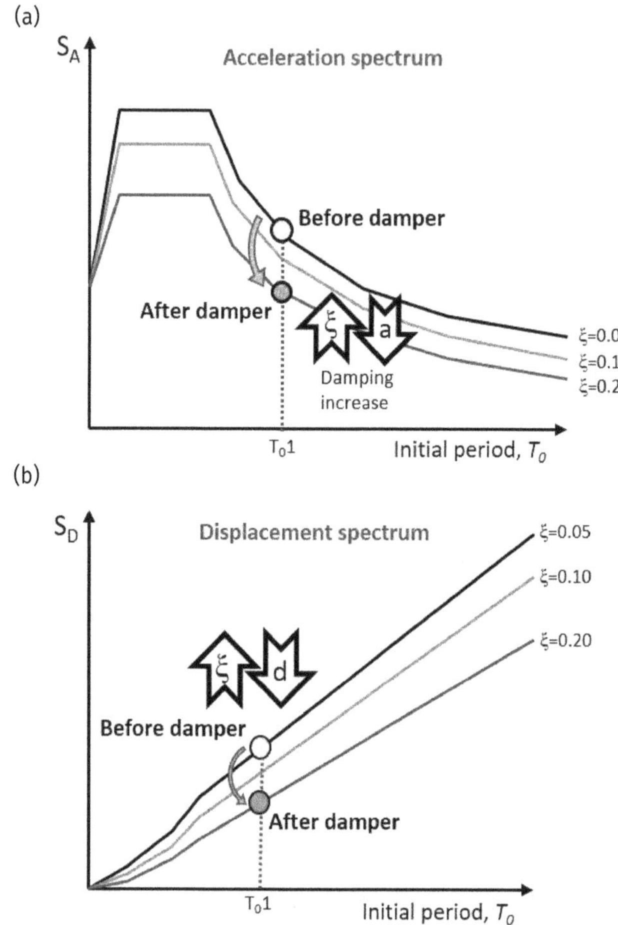

Fig. 3.34 Response control with a velocity-dependent device such as viscous damper: (a) acceleration and (b) displacement spectrum

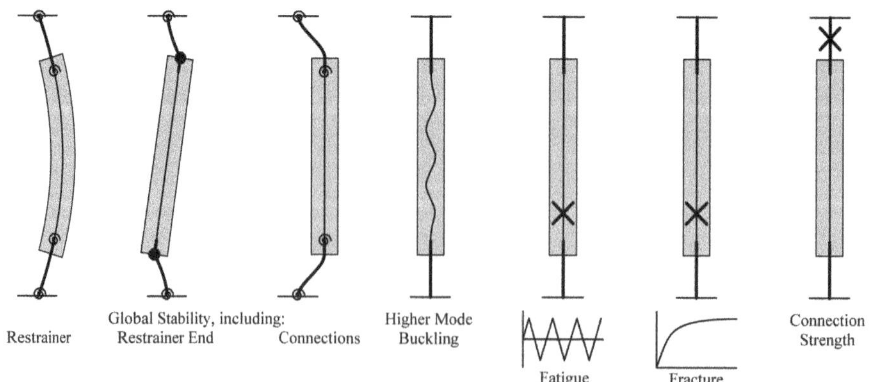

Fig. 3.35 BRB stability and strength [56]

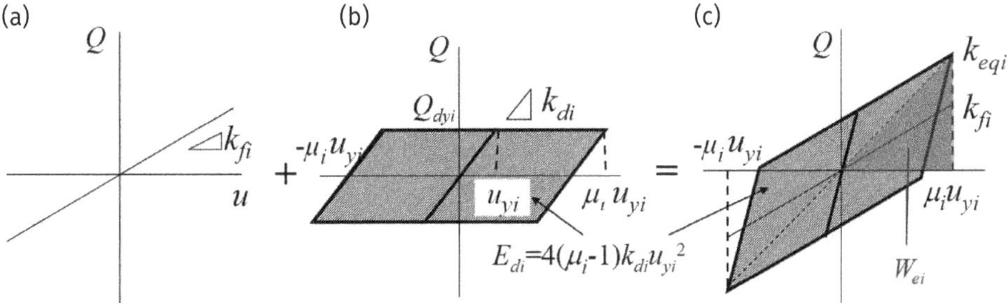

μ : Ductility ratio

k_{fi} : Frame stiffness for an elastic structure at the ith storey

k_{di} : Initial elastoplastic damper stiffness at the ith storey

k_{eqi} : Equivalent stiffness of the total system at the ith storey

u_{yi} : Yielding displacement of elastoplastic dampers at the ith storey

Fig. 3.36   Response control with elastoplastic damper: (a) elastic building, (b) elastoplastic dampers (BRB) and (c) total system

When BRBs being an elastoplastic type of damper added to an elastic building frame, the overall behaviour can be explained by using Figure 3.36, which is identical to the behaviour illustrated in Figure 3.4. As response control systems suffer seismically induced damage, it is assumed that the dampers keep the structural frame within the elastic range. This figure is similar to a seismically isolated system implementing elastoplastic dampers such as U-shape steel dampers.

3.7 Design Example Using Response Control: Environmental Energy Innovation Building, Tokyo Institute of Technology, Tokyo, Japan: Steel Low-Rise Building with BRBs

A recent example of a low-rise, eight-storey building is the Environmental Energy Innovation (EEI) Building at Tokyo Institute of Technology, which was completed in 2012 and is shown in Figure 3.37(a) and (b). The solar panel envelope structure connected to the main building and the layout of dampers are shown in Figure 3.37(c). Covered by 4650 solar cell panels, this building is designed as a self-electricity-providing building on sunny days [56], [57].

3.7.1 Objective of the Project

The EEI Building is designed to provide a total floor area of $9,000\,\mathrm{m}^2$ for research laboratories and an education centre for future energy and environmental strategies, including solar panels, fuel cells, and their applications. The building itself should adopt the concept

(a)

(b)

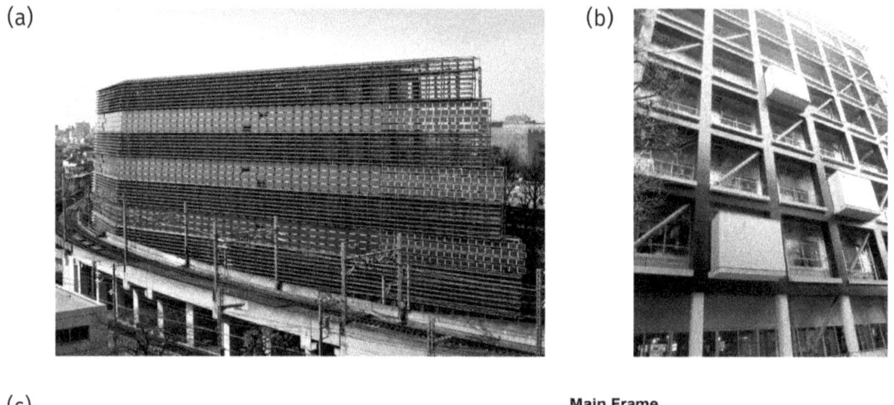

(c)

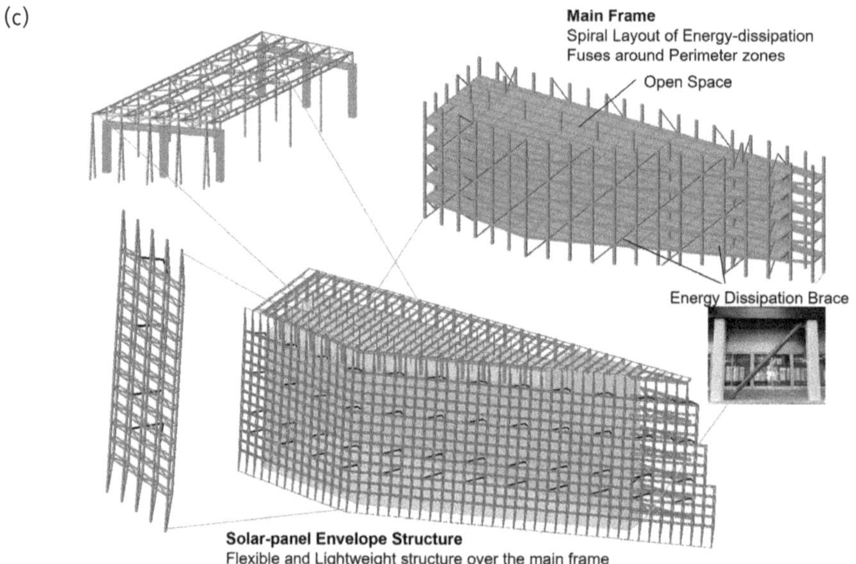

Fig. 3.37 Environmental Energy Innovation Building, Tokyo Tech, Japan: (a) south and (b) north building view and (c) structural system (Photo: T. Takeuchi)

of "Environment and Energy" by achieving high reduction ratios of energy and CO_2 emission. In the EEI Building, the drastic reduction of carbon dioxide emissions created by the research facility housed within it was a primary goal. To this end, the building utilises solar panels on three of its outer surfaces. In order to maximise the number of solar panels to be used on the long, narrow, wedge-shaped building, a detached frame of panels, the solar-panel envelope, was created.

High-performance BRBs are built into the building's outer frame as response control devices, making the EEI Building an earthquake-resistant structure while enabling the retention of large open inner spaces. This system dissipates the energy from small-scale earthquakes, reducing the response displacement and acceleration of each floor. Additionally, it avoids damage to beams, columns and the building's exterior in the event of a large-scale earthquake, thus, ensuring the long-term use of the building.

3.7.2 Design and Performance Confirmation

BRBs with low yield steel material with 225 MPa yield strength (LY225) cores are distributed along the perimeter, achieving a BRB-to-frame stiffness ratio of $K_d/K_f = 1.0$ and first storey shear strength of 10,000 kN, approximately 1/8 of the main frame shear strength. BRBs start yielding at 1/700 storey drift angle, which corresponds to a base shear ratio of 0.15, reducing the maximum storey angle to less than half (yielded BRBs are depicted with bold lines in the following figures). The results of the analyses for Level-1 (PGV=25 cm/s) and Level 2 (PGV=50 cm/s) earthquakes are presented in Figure 3.38 and Figure 3.39, respectively. While the ordinary structure without BRBs exhibit serious beam end yield and fracture under a Level-2 (10 % probability of exceedance in 50 years) earthquake (design earthquake level in Japan) (Figure 3.39), additional BRBs are sufficient to keep the main structure elastic and reduce maximum storey drift angle within immediate occupancy level. As shown in Figures 3.38(c) and 3.39(c), the dynamic response of building models with BRBs is substantially reduced compared to the ordinary frame type model in both seismic levels used in the analysis. Even in a design level event, the storey drift angles are below 1/200, which means that both structural and non-structural members will withstand this seismic level with minimum to no damage.

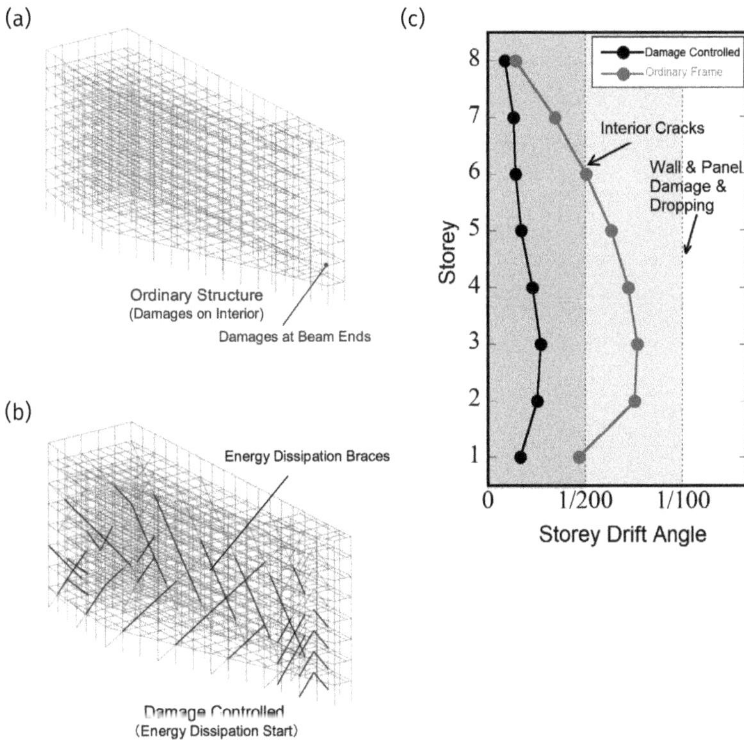

Fig. 3.38 Analyses results of damage and drifts for Level-1 earthquake (PGV=25 cm/s): (a) ordinary structure, (b) damage-controlled structure and (c) comparison of inter-storey drift [56]

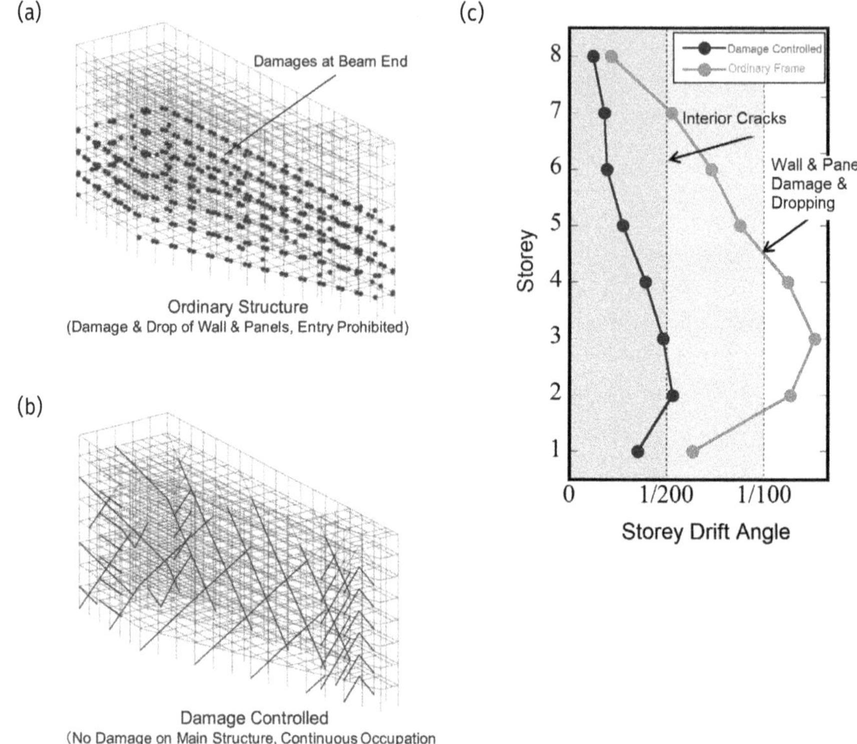

Fig. 3.39 Analyses results of damage and drifts for Level-2 earthquake (DBE) (PGV=50 cm/s): (a) ordinary structure, (b) damage-controlled structure and (c) comparison of inter-storey drift [56]

Chapter 4

Seismic Retrofit Using Seismic Isolation and Response Control

This chapter presents detailed information on seismic devices for seismic isolation and response-controlled systems available for improving the seismic performance of existing reinforced concrete structures.

Facilities such as schools and hospitals have a substantial role in civil protection in order to guarantee the continuity of the main services, especially after a major seismic event. Therefore, the continuous upgrade and compliance of these buildings to the most recent standards is of primary importance to enable the uninterrupted operation of these facilities. Recently, the development of innovative materials and subsequent advances in seismic isolation and response control systems have led to a continuous improvement of seismic protection techniques. These new methods are commonly used in structures with structural irregularities both in plan and in height, especially when they are in high seismic zones and when there are increased performance requirements.

Seismic retrofit using seismic isolation and response control systems is preferred particularly for the retrofitting of key buildings such as hospitals, governmental buildings or industrial facilities, where re-construction is not an option. These types of innovative retrofitting methods can be implemented while the building is in operation.

Many international codes provide a designated chapter for the assessment of existing buildings' structural performance and retrofitting design. However, the design of seismic isolation and response control systems is not sufficiently covered in most of these codes, and this is one of the main barriers to the implementation of these technologies. ASCE 41-17 [58], which is a code especially prepared for the seismic evaluation and retrofit of existing buildings, includes some provisions for these systems.

Retrofit design using seismic isolation or response control devices is quite similar to designing a new building with such devices. The only important difference is that the retrofit solution should be tailor-made to suit the already existing structural layout. Therefore, each retrofit project with seismic isolation or response control is a unique work that requires detailed planning and execution. In the next section, such retrofit projects are presented in a consistent format, in which the objective of each project is presented, followed by the performance requirements and information about the steps of the application.

4.1 Retrofit Design Examples Using Seismic Isolation

4.1.1 Seismic Isolation Retrofit of an RC Hospital Complex in Istanbul, Turkey

Marmara University Başıbüyük Training and Research Hospital in Istanbul encompasses 750 beds in 113.000 m² floor area and, hence, is the largest hospital in the world retrofitted with seismic isolation. The applied system consists of 688 lead-plug rubber and 154 sliding bearings [59]. The hospital building, composed of 16 independent blocks of rectangular shape rising to different heights and a block designated as a car park (Figure 4.1), was built in 1991. The building, which did not comply with the existing codes, underwent a comprehensive retrofit, as it is one of the most important hospitals in Istanbul and it is well known that the area is within a high-risk seismic zone.

The different blocks are regular both in plan and in height. However, they have different numbers of floors. Blocks A1, A2, A3 have four floors, blocks A4, A7, B4, B8 have twelve floors, blocks A8, B7 have 13 floors, blocks A5, A6 have two floors, blocks B1, B2, B3, B5, B6 have three floors, and the block for car park has two floors.

4.1.1.1 Objective of the Project

The structure is considered irregular in height due to the non-homogeneously designed blocks and also because of the level of foundation, which is not constant but shared among three different levels. The structure was retrofitted in 2002 according to the new requirements introduced by the renewed Turkish seismic standard in 1998 by adding RC shear walls and column jacketing. Nevertheless, after the code update in 2007, the recent retrofit application did not satisfy the requirements of the new standard except for the parking block. After detailed analyses and feasibility studies, it was decided to retrofit the structure using seismic isolation technology.

In a conventional retrofitting approach, the practical performance-based procedure would suggest increasing the strength by adding new ductile concrete shear walls from the inside or outside while correcting the deficiencies of the critical elements controlling the deformation capacity and stability of the existing system. However, this conventional approach has the following drawbacks:

- Construction work is required on every floor;
- Added shear walls obstruct architectural features;
- Possible damage to medical equipment and architectural elements;
- High acceleration felt during the earthquake can entail panic and disturbance;
- Medical services might be interrupted.

In contrast, seismic isolation retrofit brings the following advantages for this project:

- No interruption of medical service during and after an earthquake;
- Provides comfort through the very low acceleration transferred to the superstructure;

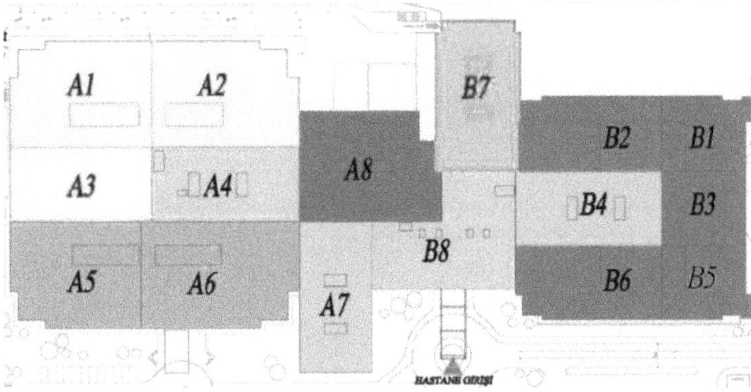

Fig. 4.1 Building layout and representation of the different blocks [59]

- No damage to neither structural and architectural elements nor medical equipment;
- Minimum obstructions to architectural features;
- No interventions at upper floors except unifying the high-rise blocks to prevent hammering.

Seismic isolation retrofit principles for this project can be listed as follows:

- Access to all possible isolator positions needs to be provided;
- Isolators will be placed below the ground level slab (minimum cost, more efficient, less complex construction work);
- Basement floors will be conventionally retrofitted by adding shear walls and jacketing the columns;
- Interventions will not be made at upper floors except unifying the high-rise blocks;
- Important partitions with vulnerable and expensive devices will be transferred from basement floors to isolated upper floors (e.g. operating theatres have been moved to 1st floor);
- All blocks will be unified in basement floors;
- Elevators and stairs will be suspended to the ground floor.

4.1.1.2 Design, Performance Confirmation and Application:

In general, in seismic isolation applications, isolation units are installed at the foundation level. However, due to the non-homogenous foundation levels of the Marmara University Başıbüyük Hospital building blocks, a seismic isolation system, also called primary isolation interface, was installed below the ground floor to ensure that the seismic response of the isolation system will not be restrained. In the retrofit layout, basement floors were conventionally retrofitted by adding shear walls and jacketing the columns. To ensure the functionality of stairs and elevators, the elevator and staircase core walls were supported with slider units (shown as "secondary isolation interface") at the foundation level, as illustrated in Figure 4.2. Sufficient gaps were implemented between the elevator/staircase core walls and the basement floor slabs to avoid collision [59].

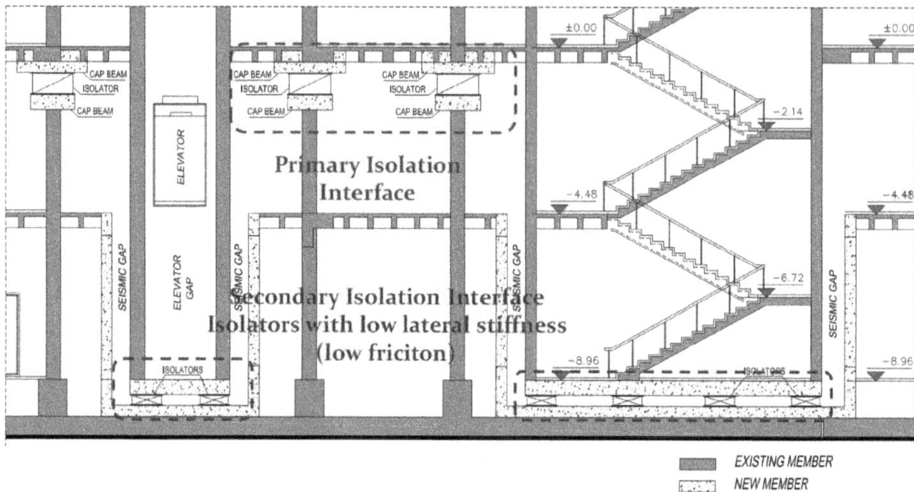

Fig. 4.2 Isolation interface location under the staircase and elevator core walls. New structural elements (light grey) are built to accommodate isolation at different levels [59]

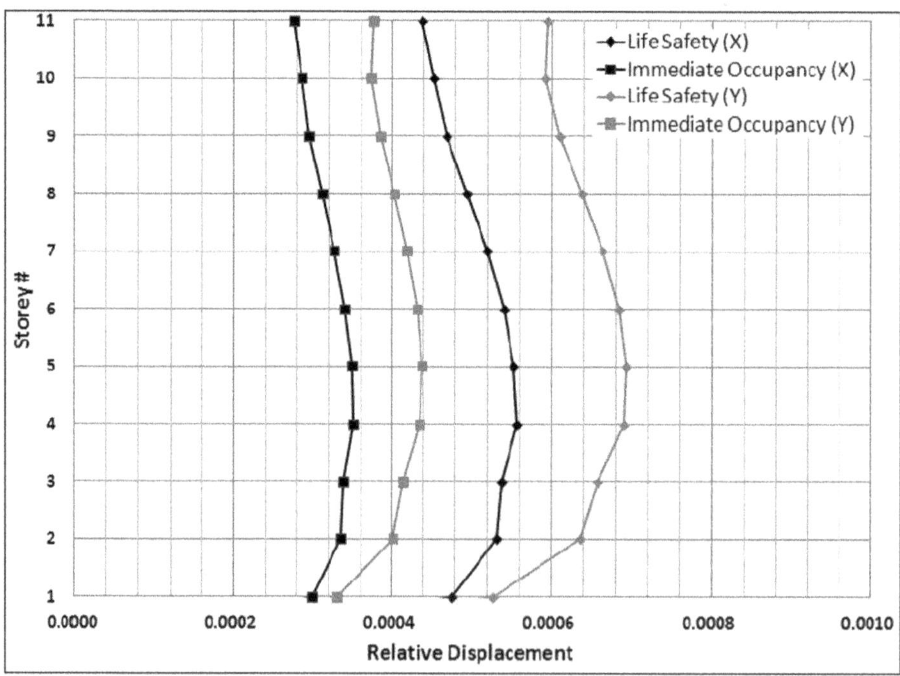

Fig. 4.3 Inter-storey-drift ratio profile for DBE (red) and MCE (blue) levels in two orthogonal directions [59]

According to the seismic isolation design targets of the project, two different structural performance levels are determined for two different seismic intensity levels. For seismic level with a 2 % probability of exceedance in 50 years (MCE level earthquake with a return period of 2475 years), the target performance level is "Life Safety"; the maximum isolator displacement has been limited to 500 mm, and the base shear transmitted to the superstructure has been limited to 13.35 % of the total seismic weight. For the design level seismic event (DBE level event with a return period of 475 years), the performance objective is to achieve the "Immediate Occupancy" level. Relative displacements (inter-storey drifts) are given in Figure 4.3 for the aforementioned ground motions and for the monobloc of higher blocks (A4, A7, A8, B4, B7, B8) as an example.

One of the most important aspects of seismic isolation retrofit using isolation units is the installation process. During the column cutting process and isolator installation, the vertical loads of the building should be temporarily and safely supported. On the other hand, vertically continuous components, such as elevator shafts or staircases, that go through the isolation layer require special attention and particular solutions.

According to [59], for the isolators' installation process in Marmara University Başıbüyük Hospital, the design team proposed an installation technology that prevents vertical deflection, which may cause damage in the building above the seismic isolation layer. The proposed technology uses two steel clamps for each column, placed with 40–50 cm distance between them. These clamps are tightened to the column through post-tensioned bars made of high strength steel. Next, by hydraulic jacks with adequate capacity, the column segment between the two clamps is unloaded. These segments are then cut with a diamond wire saw, removed, and the isolation unit is inserted in its place by grouting and anchoring the unit to the previously cut surfaces of the column. After the grout is set, the hydraulic jacks are unloaded, and this way, the vertical loads in the column are transferred to the isolator unit. The column cutting process and installed isolators are shown in Figure 4.4 and Figure 4.5. Completed retrofit work is shown in Figure 4.6.

Fig. 4.4 Jacking, diamond wire cutting and extraction of the column block (Photo: R. Turan)

Fig. 4.5   Installation of seismic isolation layer (Photo: R. Turan)

Fig. 4.6   Marmara University Başıbüyük Hospital seismic isolation retrofit application (Photo: B. Sadan)

Fig. 4.7 Marmara University Başıbüyük Training and Research Hospital after retrofit (Photo: B. Sadan)

During the column cutting and isolator installation process, all vertical deflections are constantly monitored in order to avoid post-tensioning loss and to maintain the height of the columns. A photo of the retrofitted Marmara University Başıbüyük Training and Research Hospital complex is provided in Figure 4.7.

4.1.2 Retrofit Project of a Residential Building Using Seismic Isolation, Athens, Greece

The building in question is an RC structure in Athens, Greece, representative of the residential buildings designed in the 60s and 70s in Southern Europe, thus it was designed following an older generation of codes. A characteristic transversal section and plan view is presented in Figure 4.8. A response spectrum analysis performed for the seismic assessment of the structure using EC8 code provisions showed poor structural performance and inadequate capacity of many structural members [60].

4.1.2.1 Objective of the Project

According to the updated codes, a seismic upgrade is required for this building. Instead of selecting conventional techniques such as application of RC jackets, fibre-reinforced polymers (FRP) or implementation of new RC shear walls or steel braces, seismic isolation is adopted. To this end, Friction-Pendulum System (FPS) bearings are placed at the foundation level.

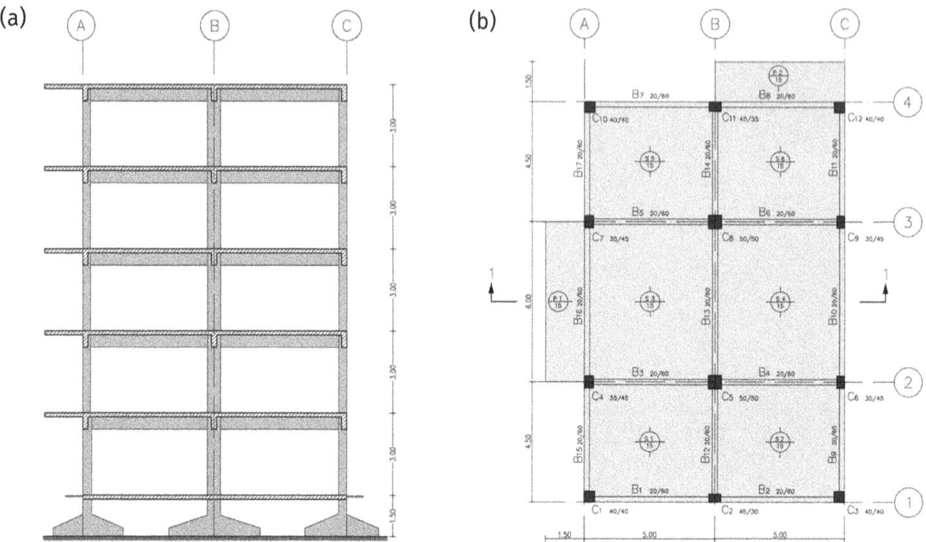

Fig. 4.8 (a) Transversal section 1-1 of the building and (b) ground floor layout

4.1.2.2 Design and Performance Confirmation

For the assessment and the seismic upgrade of the building, the elastic response spectrum of EC8-1 [42] is used. Athens has a reference peak ground design acceleration of a_{gR}=0.16 g, the soil factor, S, is equal to 1.2 for soil category B, the building importance factor is γ_I = 1.0 corresponding to the importance class II (residential use) and the damping factor η = 1 (for viscous damping, ξ=5 %). Performance level B1 is considered, as defined in EC8-3 [61]. This corresponds to a requirement of performance of life safety level with a possibility of exceedance of 10 % in 50 years for the seismic action, i.e. an average return period of approximately 475 years. The elastic response spectrum is depicted in Figure 4.9 as "Elastic design". In the same figure, the dashed line depicts the equivalent constant response spectrum corresponding to the 1959 Greek Seismic Code originally used for the design of the structure.

For the seismically isolated structure, the spectra for the design of the superstructure, the foundation and the isolation system are derived from the elastic response spectrum. In the horizontal direction, for the design of the superstructure and the isolation system, a damping correction factor of η=0.70 is applied to the elastic design spectrum. Regarding the behaviour factor, a value of q =1.5 is used for the superstructure design response spectrum corresponding to quasi-elastic behaviour. For the isolation system, no behaviour factor is applied (q=1) when checking the stress condition. Furthermore, a magnification factor, γ_x, as required by the code, is applied to the displacements. The value of γ_x used is 1.5, as ascribed by the Greek National Annex of EC8-1.

The alternative approach is the use of seismic isolation at the foundation level. The behaviour of the structure is then tested through a series of linear and non-linear analyses. The results indicate that all the existing structural members above the isolation level have adequate capacity. In addition, inter-storey drifts are significantly lower than the ones

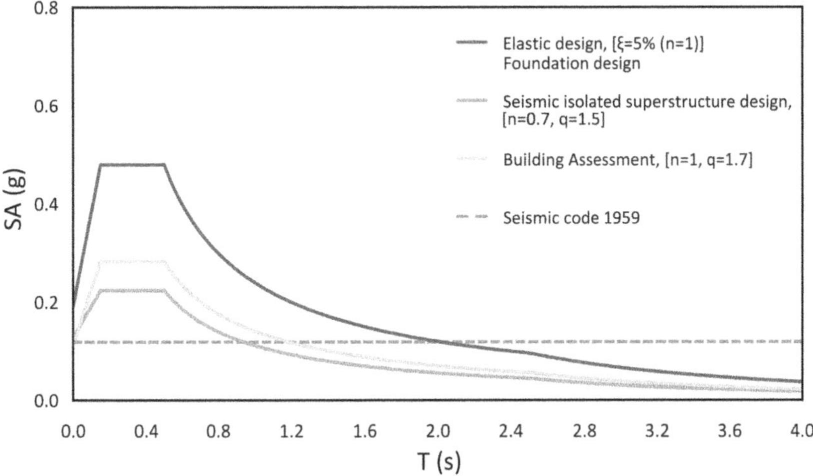

Fig. 4.9   Horizontal acceleration response spectra

Fig. 4.10   Structural model of the building before the retrofit. The red diagonal lines represent the masonry infill walls

of the original structure but also lower than those of the conventionally strengthened one. Furthermore, in contrast to a conventional seismic upgrade approach, no structural interventions are required above the seismic isolation level. This means that the original architectural concept can be preserved, and the seismic upgrade can take place in a shorter time and in a smooth fashion for the residents.

The inter-storey drift ratios for the original building, the conventionally retrofitted one and the seismically isolated one, are presented in Table 4-1. It can be concluded that inter-storey drift ratios for the seismically isolated structure are approximately at 25 % of the ones before the intervention and considerably lower than those of the conventionally retrofitted structure.

In order to implement the seismic isolation system, a diaphragm above and below the isolation level is required (Figure 4.11(a)). A rigid mat foundation encasing the existing spread footings through the use of dowels serves as a diaphragm below the bearings, while a system of RC beams and steel beams (Figure 4.11(b)) constructed at the ground floor level functions as a diaphragm above. It should be mentioned that in order to provide access to the isolators for future inspections and possible replacements, the ground floor level consists of the aforementioned RC beams and a steel structure supporting a removable flooring system (composed of structural steel grid and marble or wood) as shown in Figure 4.11(b).

There are two basic requirements for the implementation of the isolation system for the specific building: (a) a minimum height of 1.5 m from the upper part of the ground floor slab to the foundation level is necessary to provide working space for the construction works; if this is not available, then underpinning the foundation could be an option. However, this would increase the cost; (b) the building should be detached, which is important for the creation of the seismic joint but also to provide working access from the sides during construction.

Regarding the construction sequence, this can be done in three stages: (a) In the first stage, the existing spread footings are encased into a mat foundation, and the new RC beams at the ground floor level are constructed, thus, creating the rigid diaphragms below and above the isolator level. Also, a retaining wall at the perimeter of the building, from the foundation to the ground level, is raised for the seismic joint to be created. Particular care is required to ensure the free movement of the structure at the seismic joint. If the base isolation system is unable to move freely, the seismic performance of the base-isolated superstructure would be significantly compromised. It is, therefore, not permissible to restrict the movement at the joint. (b) In the second stage, the columns between the two diaphragms are cut, and the bearings are implemented using a system of hydraulic jacks. This work could be carried out in 3 sub-stages corresponding to 3 groups of four columns, each treated simultaneously, considering safety and cost. (c) In the third stage, steel beams are connected to the sides of the RC beams. These steel beams support the flooring composed of a structural steel grid and a marble or wood cover. The flooring is removable, providing access to the bearings for inspection and replacement.

Table 4.1   Inter-storey drift ratio, γ, for each floor in the longitudinal, x, and the transversal, y, direction

Storey	Existing structure (assessment) RS analysis		Retrofitted structure shear walls & jackets RS analysis		Seismic isolated structure RS analysis		Seismic isolated structure t-h analysis	
	$\gamma_{x,max}$ (‰)	$\gamma_{y,max}$ (‰)	$\gamma_{x,max}$ (‰)	$\gamma_{y,max}$ (‰)	$\gamma_{x,max}$ (‰)	$\gamma_{y,max}$ (‰)	$\gamma_{x,max}$ (‰)	$\gamma_{y,max}$ (‰)
5-4	1.59	1.44	1.10	0.92	0.41	0.38	0.39	0.33
4-3	2.75	2.36	1.17	0.98	0.53	0.49	0.54	0.49
3-2	3.59	3.00	1.16	0.98	0.73	0.66	0.68	0.61
2-1	4.16	3.39	0.96	0.82	0.91	0.82	0.78	0.71
1-0	3.15	2.67	0.47	0.42	0.89	0.79	0.71	0.67

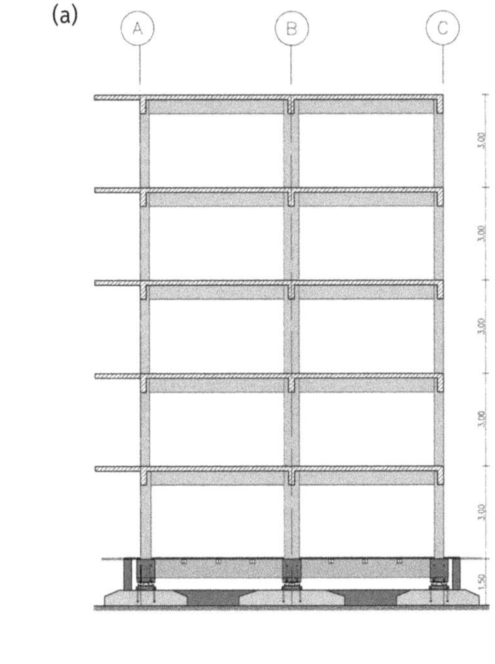

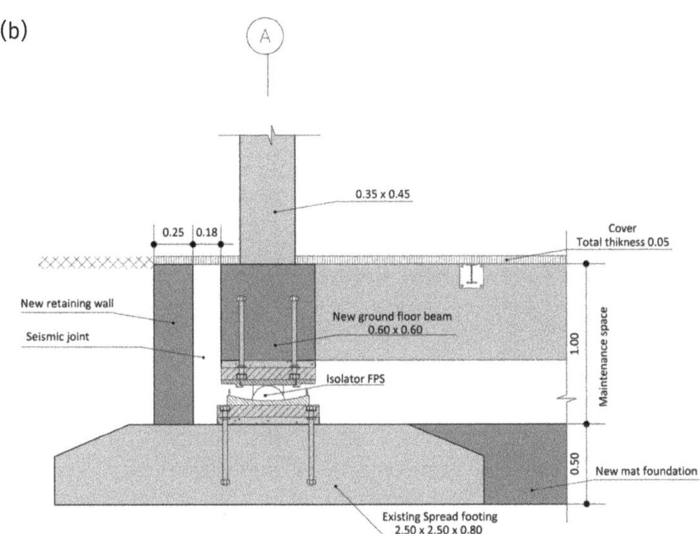

Fig. 4.11 Structural interventions for the implementation of seismic isolation (a) transversal section of the building (b) section detail. Dark shaded areas denote the interventions.

A cost comparison was made between the conventional and the seismically isolated retrofit. It was found that the latter would be realised at 18 % of the cost of a new building (in Greece), while the former at 24 % of the cost. This comparison does not consider the cost of the residents' relocation during the intervention works or the cost of the used-up space by the new walls and jackets of the conventional retrofitting. Furthermore, the added value of the seismic isolation is enormous. This comes as a result of the enhanced performance of a seismically isolated structure compared to a conventional one and the preservation of the original architectural form of the structure. Additionally, since no interventions are required on the upper floors, the seismic upgrade will take place not only in a much shorter time but also in a smooth way with minimum nuisance for the residents. Finally, regarding construction time, 12 weeks are estimated for the conventional retrofitting versus five weeks for the implementation of the seismic isolation technique.

4.2 Retrofit Design Examples Using Response Control

4.2.1 Retrofit of an RC Building Using BRBs Including an Integrated Façade, Tokyo, Japan

The Midorigaoka-1st building at Tokyo Institute of Technology is a 6-storey RC building designed in 1966 prior to the revision of the Building Code of Japan in 1971. The elevation and the plan of the building are shown in Figure 4.12. The building has a 23.3 m × 60 m rectangular plan with cores for stairs and restrooms. Horizontal stability for the N-S direction is secured by strong RC walls; however, the structural system for the E-W direction is composed of moment frames with fragile columns and beams with 4.0 m spans. The values of the seismic capacity index I_S [62], which indicates the seismic capacity according to current standards in the E-W direction, are much less than the target value of 0.7, which is the minimum acceptable seismic capacity index level according to the Japanese Seismic Code [44]. With a minimum seismic index value of 0.27 obtained on the second floor, there is a high risk of a soft storey collapse.

4.2.1.1 Objective of the Retrofit Project

Elasto-plastic energy-dissipation devices as BRBs have several desirable characteristics that frequently receive attention for retrofit projects. In brace configurations, BRBs are relatively stiff and can be designed to yield at small drifts, providing substantial energy dissipation. As the compressive and tensile strengths are nearly equal, connection overstrength demands are minimised, reducing the required amount of local strengthening. It is also possible to insert BRBs strategically to act as fuses, protecting particular members.

One of the typical retrofit strategies for non-ductile moment frames is to install BRBs as bracing elements along the perimeter, either as an additional external frame or in-plane with the existing frame. However, such retrofits are often difficult to be implemented while maintaining the continuous occupation and, frequently, have a negative effect on the building's aesthetics. At this point, it should be recognised that façades have various functions. They are not only a suitable location for seismic reinforcement but also affect energy efficiency and architectural appearance. To resolve these competing functions, the concept of "Integrated Façade" can be employed, treating structural retrofit, façade design and environmental design combined together, including improvements on seismic performances using seismic energy dissipation devices.

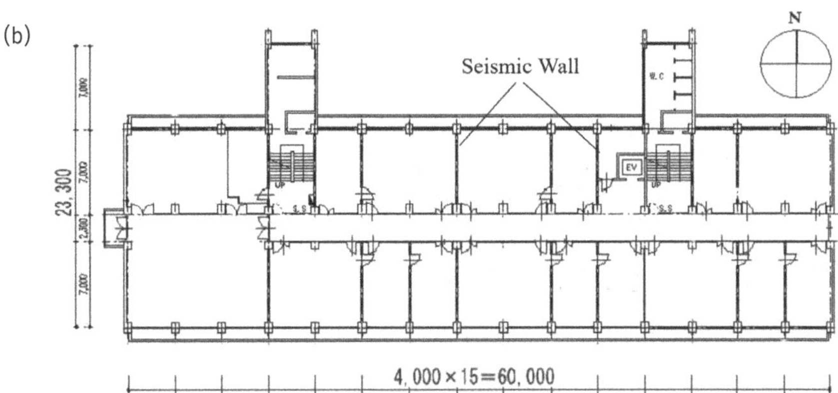

Fig. 4.12 (a) Building elevation and (b) plan view before retrofit (Photo: T. Takeuchi)

An example of an Integrated Façade concept is shown in Figure 4.13 and Figure 4.14. The existing structure is retained to support gravity loads and provide elastic stiffness, while BRBs are introduced outside of the glass façade, serving both as seismic energy dissipation members and as attachment points for an additional outer skin of louvres or glass.

The façade is not designed as a pure structural member or architectural element but integrates these two functions. Structurally, either BRBs or other dampers could be used for the brace elements, dissipating energy, reducing the response and protecting the main structure, as shown in Figure 4.13. Environmentally, the double skin is provided to improve the energy efficiency of the glass shell (Figure 4.15 (a) and (b)), which is popular in Europe. However, its effects are focused on winter and require an air ventilation system in summer. In subtropical areas, such as Japan, outer skin systems composed of louvres, as shown in Figure 4.15 (c) and (d), can be employed.

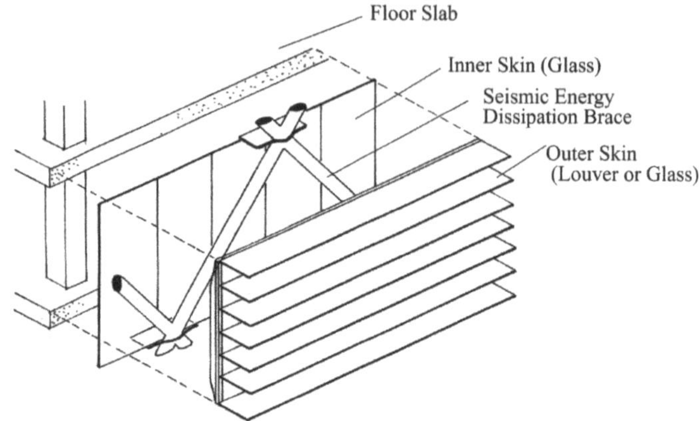

Fig. 4.13   Concept of integrated façade [56]

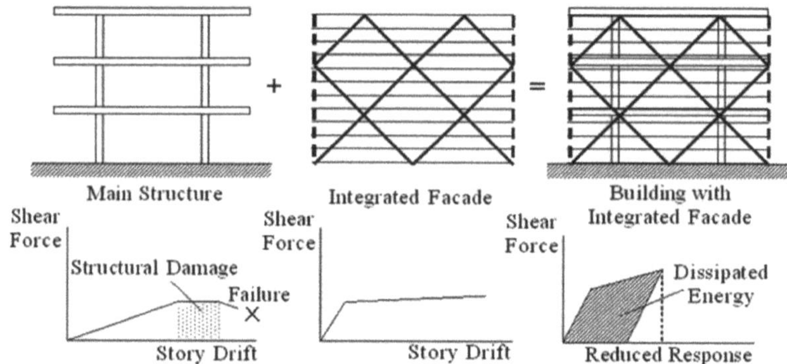

Fig. 4.14   Concept of the structural design of an integrated façade [63]

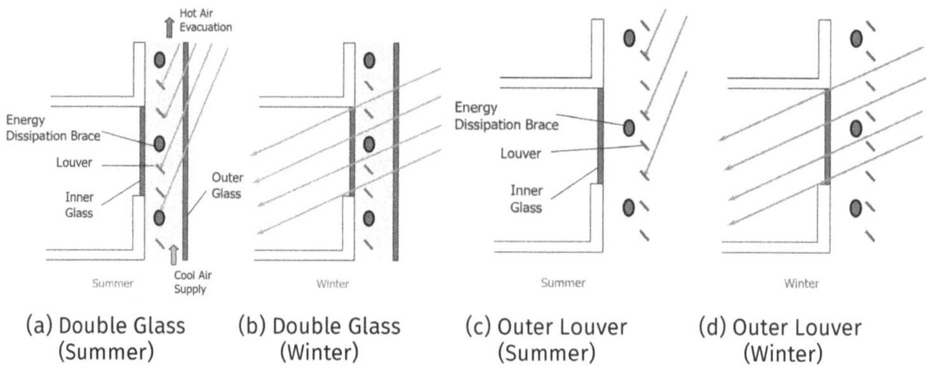

(a) Double Glass   (b) Double Glass   (c) Outer Louver   (d) Outer Louver
   (Summer)   (Winter)   (Summer)   (Winter)

Fig. 4.15   Concept of physical environment control design [63]

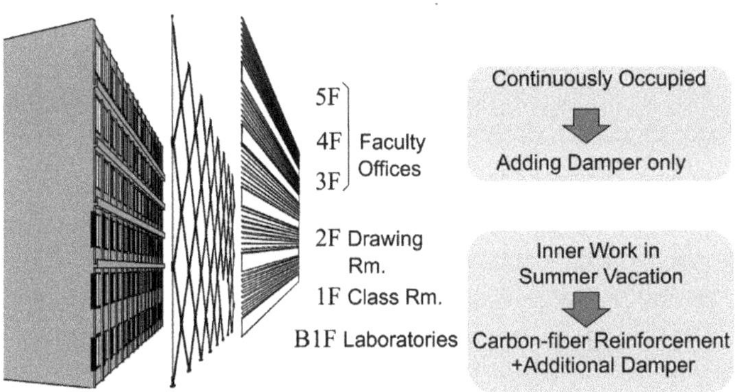

Fig. 4.16 Integrated façade retrofit concept [63]

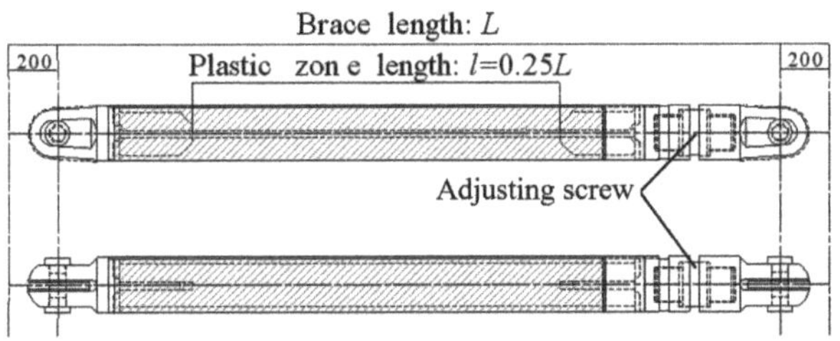

	Mark	Size	Strength (kN)	No.	Position
	RB1	PL-19×152 (LY225)	650	15	4F
BRB	RB3	PL-32×153 (LY225)	1102	13	3-4F
	RB4	PL-40×159 (LY225)	1431	80	B1-3F

Fig. 4.17 Applied BRB members [63]

The proposed integrated façade concept was applied to the Midorigaoka-1st building [56]. The retrofit concept is shown in Figure 4.16. The lower floors (B1F-2F) of the building are classrooms, which are not used in summer, so that replacements of window frames and column reinforcement are possible then. Columns in these floors are reinforced by carbon fibre sheets, preventing shear failure and improving deformation capacities. To achieve the target strength, additional BRBs with louvres are attached to the façades. These braces are designed to yield at the required minimum strength and then start dissipating energy as hysteretic dampers.

A standard BRB design with circular restrainers and cast pin connections was applied, as shown in Figure 4.17, but a special adjustable detail (adjustable screw) was also installed to improve permissible site tolerances. In Figure 4.17, a brief table summarises the type, size, number and position of the applied BRBs in the building.

4.2.1.2 Design, Performance Confirmation and Application

For evaluating the seismic performance of the retrofit, three types of BRB (Figure 4.17) with plastic zone length ratios of l/L = 0.7, 0.4, and 0.25 were examined. Also, ordinary

H-section braces with buckling strengths equivalent to the BRB yield strength were considered. Each alternative was added to the shear force-storey drift relationship of the RC main frame. For the evaluation of the hysteresis of the main frame, the following three types of ultimate strength indexes on E-W direction were compared.

(A) Ultimate storey strength determined from the seismic index Is, which is calculated for each vertical structural member (column and shear walls) at each storey according to retrofit design recommendation for RC frames [56]. The corresponding displacements are determined according to the member ductility, with the critical member, in this case, achieving the ultimate capacity at an inter-storey drift ratio of 1/250.

(B) Summation of the column and shear wall strengths at cracking and ultimate shear stages according to the Architectural Institute of Japan (AIJ) RC design recommendations. The initial stiffness and cracked stiffness are also provided, giving the total storey strength and displacements.

(C) 3D pushover analysis, with member cracking and ultimate strengths determined from method (B) but including column axial and beam flexural deformations for more precise evaluation.

The obtained shear force-deflection curves for the 2nd floor before retrofit according to each method are shown in Figure 4.18(a). The actual shear force-storey displacement relationship is assumed to follow the line (C) until 15 mm displacement (the equivalent of 1/250 storey drift angle as the storey height is 3750 mm), and then in this design approach, it is assumed to deteriorate along the line (A). Shear force-displacement curves on the second floor after retrofit calculated by the methods (A)–(C) are shown in Figure 4.18(b). The relationship between calculation methods is similar to Figure 4.18(a). However, the ultimate strength values are almost twice as high, and in this case, the building will not exceed the 1/250 storey drift angle, and therefore line (A) will not be used. The effect of the plastic zone length of BRBs is observed on the difference of yield drifts.

For verifying the evaluated hysteretic characteristics of frames, reduced mock-ups for the second-floor frame, before and after retrofit, were fabricated, and cyclic loading tests were carried out up to 1/50 storey drift angle. The test results are shown in Figure 4.19. The 1/2.5 scale specimens, which represents the second-floor frames, have axial force introduced by post-tensioning bars. The obtained hysteretic loops are shown in Figure 4.19 (a) before and (b) after retrofitting. In the un-retrofitted frame, shear failure started at 1/150 storey drift angle ($\delta = 10$ mm), then horizontal capacity drastically degraded by each cycle (Figure 4.19 (a)). In contrast, the frame after retrofit, which was reinforced by carbon fibre and BRB, did not present shear failure but showed stable hysteretic loops up to cycles of 2 % storey drift angle (Figure 4.19 (b)).

Time-history analyses were carried out with the hysteretic loop models shown with dashed lines under the condition of $l/L = 0.25$. Seismic waves of El Centro NS, Taft EW, Hachinohe NS, and JMA Kobe NS with PGV scaled to 50 cm/s (Level-2), and the design artificial wave BCJ-L2 were used for the 3-D analytical models of the building. Figure 4.20(a) shows the response of the building before retrofit, whose maximum drift exceeds 1.0 %. Considering the results of the mock-up test, the building is expected to collapse

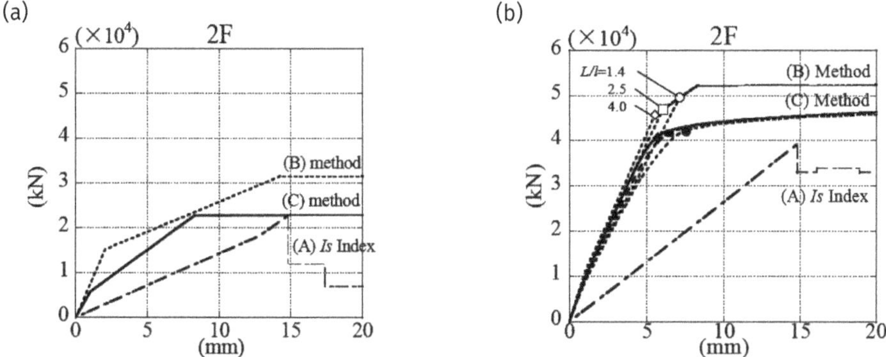

Fig. 4.18 Shear force-storey displacement relationship at second Floor: (a) before and (b) after retrofit [63]

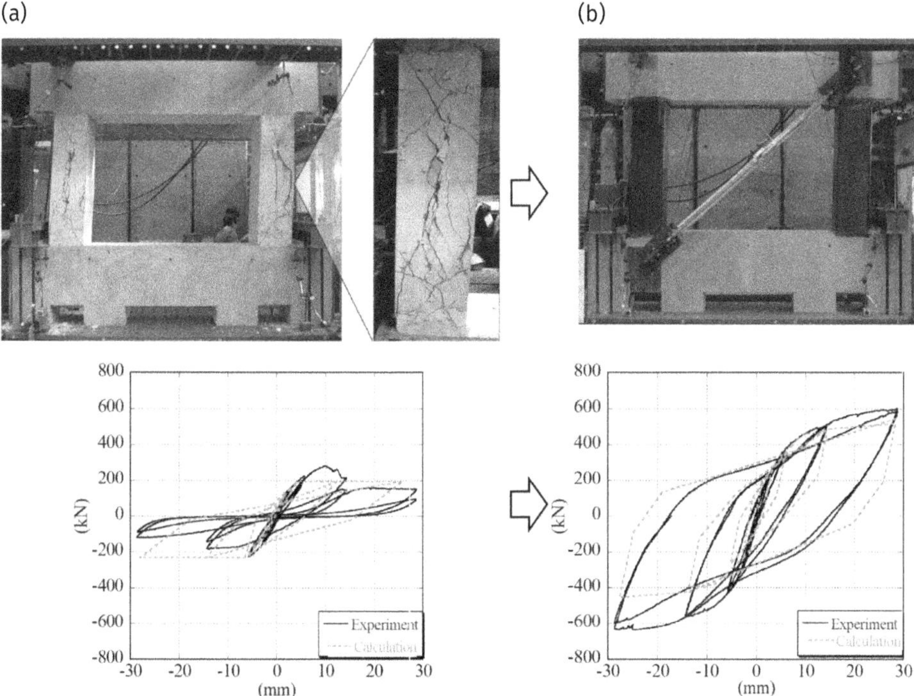

Fig. 4.19 Reduced mock-up test for the second-floor frame: (a) before and (b) after retrofit [63]

with the shear failure at the second-floor columns in a real situation. Figure 4.20(b) shows the response with BRB with plastic zone length ratios of $l/L = 0.25$, which satisfies the maximum storey drift being less than 0.4 %, which means almost no damage to the main structure.

The connection between the BRBs and the existing building transfers a horizontal force of up to 2800 kN and was detailed to avoid obstructing the building functionality. A typical detail is shown in Figure 4.21. First, chemical anchors are drilled into the perim-

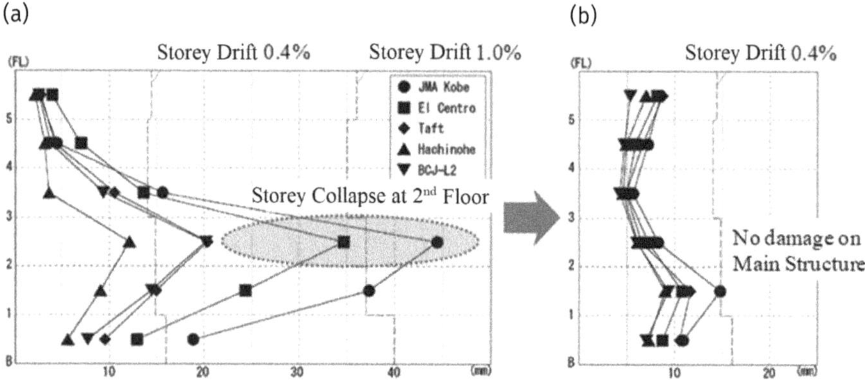

Fig. 4.20   Storey drift: (a) before and (b) after retrofit [63]

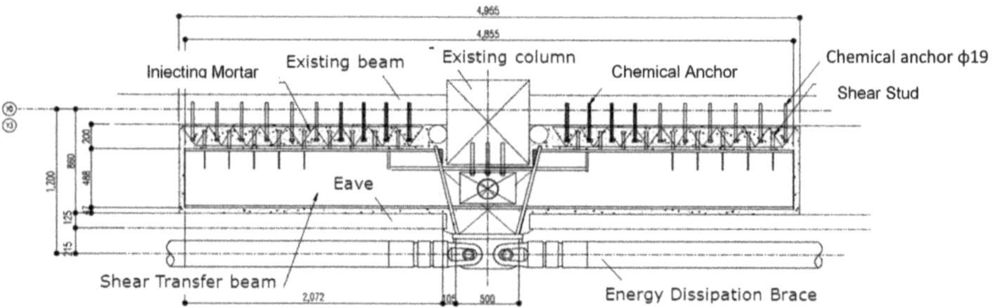

Fig. 4.21   Detail of the connection between frame and BRB [63]

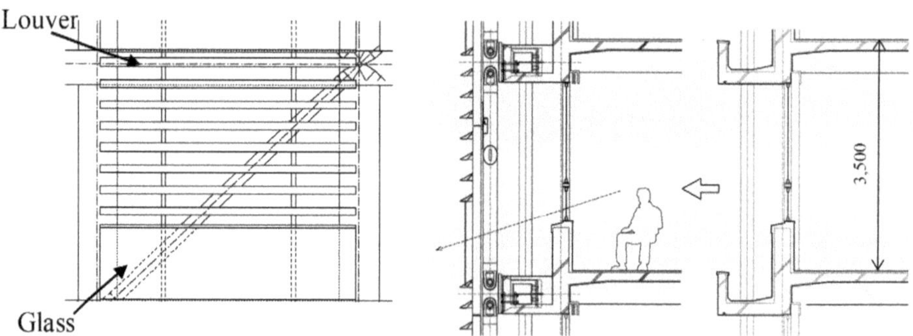

Fig. 4.22   Detail of the outer skin [63]

eter beams from the outside. Next, steel beams with shear studs are inserted in eaves and fixed to perimeter beams by injecting mortar. Finally, the BRBs are connected to brackets attached to the outer face of the shear transfer beams.

The thermal performance of the new façade (Figure 4.22 and Figure 4.23) was also investigated using Computational Fluid Dynamics (CFD) analysis, with the louvres proving effective in providing shade from the summer sun while acting as light shelves to improve daylighting and as double skin to improve energy efficiency in the winter.

(a)   (b)   (c)

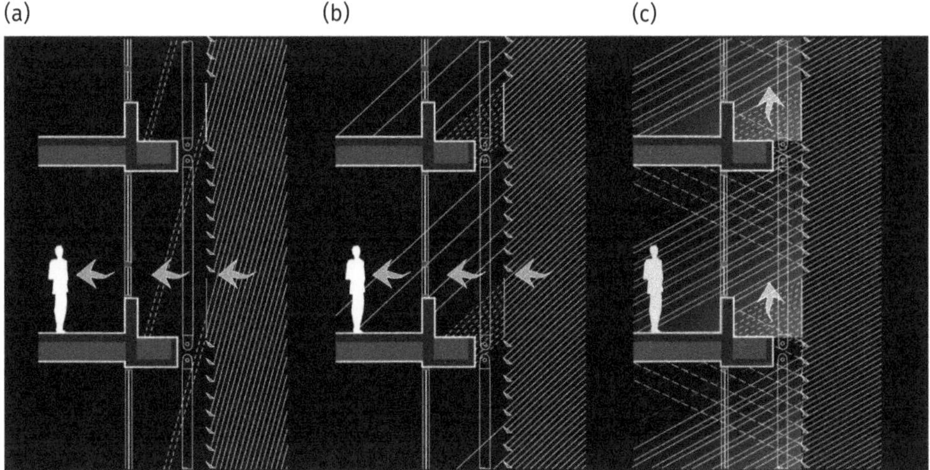

Fig. 4.23 Environmental effect of outer skin: (a) Summer, (b) Spring/Autumn and (c) Winter [63]

(a)   (b)   (c)   (d)

Fig. 4.24 Perimeter work process: (a) original eave, (b) setting steel beam, (c) attachment of BRB and (d) louvre and glass (Photo: T. Takeuchi) [63]

The construction sequence is shown in Figure 4.24. It started in July 2005 and was completed the following March 2006. Louder construction activities were carried out in the second month during the university holiday, with minimal disturbance to occupants. In Figure 4.24, the process of retrofitting at the perimeter zone is shown. Figure 4.24(a) shows the original eave, while Figure 4.24(b) the setting of the connection beams in the eave. In Figure 4.24(c), the attachment of BRBs on the concrete frame is presented, followed by the attachment of outer skin on BRBs in Figure 4.24(d). Prior to the brace attachment (Figure 4.25(b)), carbon fibre reinforcement for the columns has been completed (Figure 4.25(a)). Figure 4.26 shows the building in elevation after the completion of the retrofitting process, which represents a completely renewed appearance. The retrofit work was carried out within the ordinary retrofit cost range and schedule even with the outer skin because no additional piles were required.

(a) (b)

Fig. 4.25 (a) Carbon-fibre reinforcement, before and after application and (b) BRB attachment (Photo: T. Takeuchi)

(a) (b)

Fig. 4.26 Building façade after retrofit: (a) full view and (b) detail

4.2.2 Full-Scale Tests on Response Control Retrofit for an RC School Building Using BRBs, Istanbul, Turkey

Response controlled retrofit, implementing BRBs and elastic steel frames was designed and experimentally confirmed for a typical sub-standard five-storey RC school building in Turkey (Figure 4.27). The school building was constructed in 1992 in Istanbul, which is a high-risk seismic zone. The average concrete compressive strength was 20 MPa, and structural member cross-sections, as well as the steel reinforcement, do not satisfy the current design code of Turkey [47]. Storey height for each storey is H_1=3.2 m. In this study, the investigation of the building and evaluation of the structural performance before and

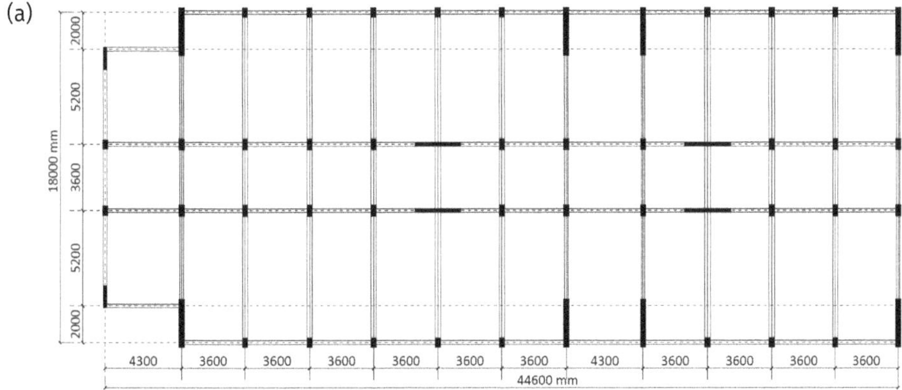

Fig. 4.27   RC school building in Turkey: (a) plan and (b) elevation view (Photo: F. Sutcu)

after the BRB retrofitting method were carried out in the longitudinal direction only, which is the weaker direction [64].

4.2.2.1 Objective of the Project

In order to apply energy-dissipation devices to retrofit reinforced concrete structures, careful design of the connections is required. For example, a challenging task is to achieve an in-plane attachment of BRBs on an existing RC frame in a practical manner. One of the most commonly recommended configurations concerns inserting steel frames with energy-dissipation devices within the existing RC frame, as shown in Figure 4.28. This is achieved by attaching post-fixed chemical anchors to the RC frame and by welding shear studs to the steel frame, using mortar infill and stirrups to establish continuity. The Japanese Standard for Seismic Diagnosis of Existing Reinforced Concrete Structures [65]

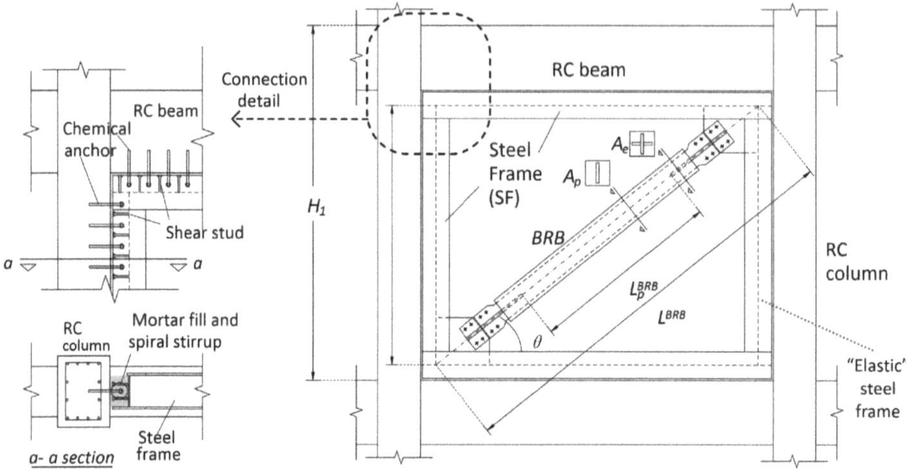

Fig. 4.28 Layout of BRB retrofit application on an RC frame with connection details [64]

recommends this connection to be designed for the combined strength of the steel frame, including the energy dissipation devices. With the proposed application, all retrofitting works were conducted outside the building, and, this way, the building could be kept operational with minimum interruption of its educational purpose.

In more detail, the proposed BRB retrofitting system includes a supplementary Steel Frame (SF) that is designed to remain elastic, installed in parallel with the BRB, in which the SF enhances the system stiffness and restoring capability. All BRBs employed a low yield point steel core (LYP225) and were either welded or bolted to gusset plates integral to the SF. The connection between the RC and steel frames consists of steel studs, chemical anchors, ladder stirrups, and high-strength grout. This connection interface, denoted as the mortar zone, provides significant composite action, which is clearly identified in the test results that followed.

The proposed retrofit application is expected to reduce the residual inter-storey drift ratios, enhance self-centring features, and allow repairability of the building after a major seismic event. Moreover, the SF increases the axial shear and bending capacities of the involved RC columns, which is especially important in similar applications where the addition of the BRB members would increase the seismic demands on the columns.

4.2.2.2 Design, performance confirmation and application

The target inter-storey drift ratio for a retrofit project is specified in various codes. In NEHRP Commentary on the Guidelines for Rehabilitation of Buildings (FEMA 273) [66], as well as FEMA 356 [67] and ATC-40 [68], the maximum inter-storey drift ratio to achieve immediate occupancy is limited to 1/100. However, the Japanese Standard for Seismic Diagnosis of Existing Reinforced Concrete Structures [65] recommends 1/150, which is also used in this case study.

The design steps for the response control retrofitting scheme of the school building are described in detail in [59]. Push-over analysis curves of the existing RC frame are shown

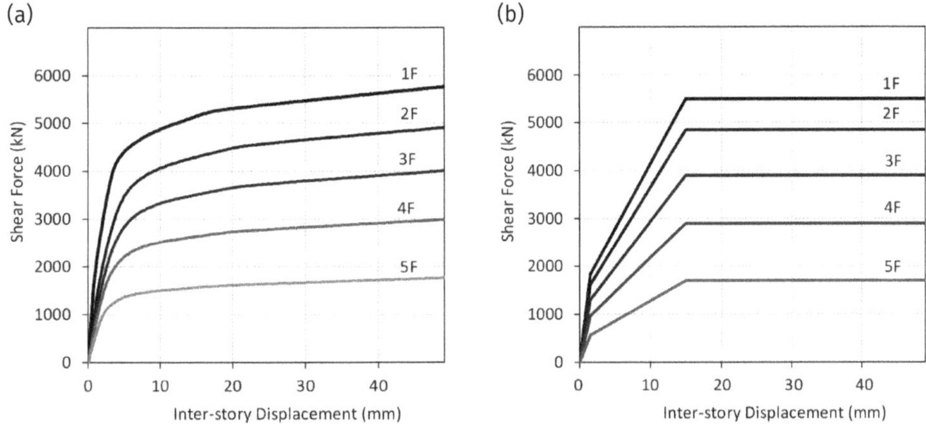

Fig. 4.29   (a) Push-over curves for the model (b) simplified tri-linear behaviour [64]

in Figure 4.29(a), which are simplified through their idealisation as trilinear models (Figure 4.29(b)) based on the equivalent energy. The eigenvalue analysis indicates an initial period of T_0 = 0.7 sec, with the equivalent mass of M_{eq} = 4433.5 tons and height of H_{eq} = 10.5 m. The initial equivalent stiffness is K_f = 332.8 kN/mm. The frame ductility (μ_f) at the target storey drift ratio of 1/150 is obtained as 1.42.

BRBs were selected with a core area of A_c=55.5 cm², yield strength of 225 N/mm², plastic length ratio of L_p/L_0=0.5 and elastic area ratio of A_e/A_c = 2.5 (see Figure 4.28 and Table 4-2). The BRBs were installed at an angle of θ = 36.4° in a steel frame consisting of W10×10×49 (H-250×250×9×14) steel column and beam cross-sections with a yield strength of 325 N/mm², producing a steel frame to BRB stiffness ratio of 0.047. The damper to frame stiffness is calculated as the initial stiffness of dampers against horizontal stiffness of the frame K_d/K_f= 2.25, corresponding to the BRBs and the steel frame system resisting 30 % of the base shear. The equivalent damping of the retrofitted system is then calculated as ξ_{eq} = 0.22, with an equivalent period of $T_{\Sigma\mu}$ = 0.81 sec.

The design spectrum is shown in Figure 4.30, with the retrofitted frame resulting in a spectral displacement of 6.69 cm, corresponding to a 1/157 inter-storey drift ratio and achieving the target storey drift. The number of BRBs installed at each level n_{di} is then calculated from Eq. (4.1).

$$n_{di} = \frac{K_{di} \cdot \delta_{dyi}}{Q_{dy1}} \tag{4.1}$$

Where

$\quad K_{di}$: the stiffness of a single BRB

$\quad Q_{dy1}$: the yield force of a single BRB

$\quad \delta_{dyi}$: yield displacement, which can be evaluated by scaling from the SDOF model by $\delta_{dyi}=(H_1/H_{eq})\,\delta_{dy}$.

For comparison, a retrofit alternative employing conventional braces (CBs) was also assessed. Conventional brace cross-sections are designed using the same material as the

(a)

(b)

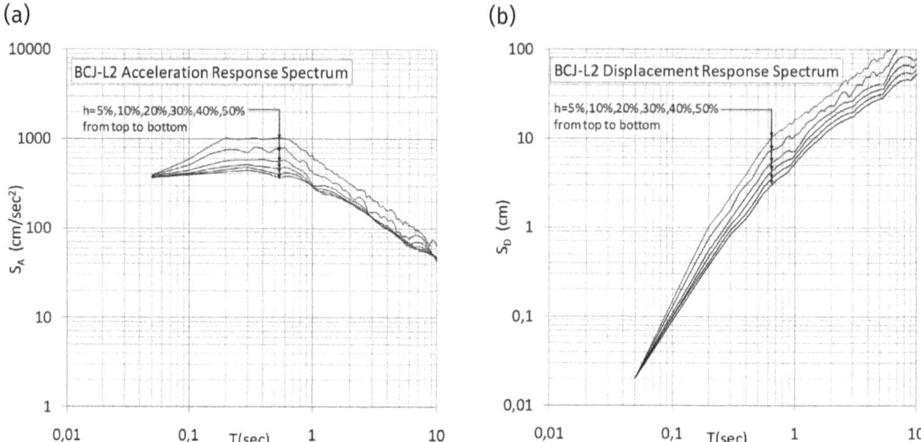

Fig. 4.30   (a) Acceleration and (b) displacement design response spectrum (h=5 %, 10 %, 20 %, 30 %, 40 %, 50 %) [64]

BRBs. The member lists of BRBs and CBs are shown in Table 4-2. A time-history analysis of lumped mass models was then conducted for the bare RC frame, BRBs without the steel frame and two retrofit schemes (Figure 4.31), with the CBs modelled using the post-buckling hysteresis model proposed by Shibata-Wakabayashi [69].

The maximum storey drifts obtained by the Level 2 design basis earthquake (BCJ-L2), as defined by the Japanese Seismic Code [44], are shown in Figure 4.31. The bare RC frame greatly exceeds the target drift (solid line with circular points), while the BRB retrofit (solid thick line with rhomboidal points) controls the response within the acceptable limits of 1/150. However, when the steel frame is excluded, damage concentrates at the second storey and drift at this floor is obtained slightly over the target displacement (solid line with triangular points). Due to the unbalanced behaviour of the CBs in compression and tension, the damage tends to concentrate on the first storey once buckling is initiated (dashed line with circular points).

Table 4.2   Number of BRBs and CBs applied per floor(F)

	BRB cross-section	K_{di} (kN/mm)	Number of BRBs	CB cross-section	Number of CBs
5F	Plastic core 222mm×26mm A_p=5550mm²	850	4	W8×8×35 A=6540mm²	4
4F		1450	6		6
3F	Steel restrainer case	1950	8		8
2F	Unbonding material	2425	9		9
1F	Mortar infill	2750	11		11

In the proposed retrofit, the steel frame is intended to remain elastic, providing significant restoring force and reducing residual displacements. The displacement time history is shown in Figure 4.32 for the bare and retrofitted frames, both with and without the

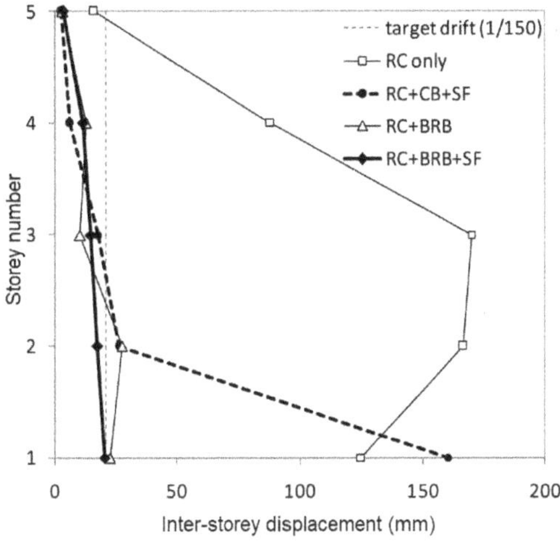

Figure 4.31 Maximum inter-storey displacement distribution [64]

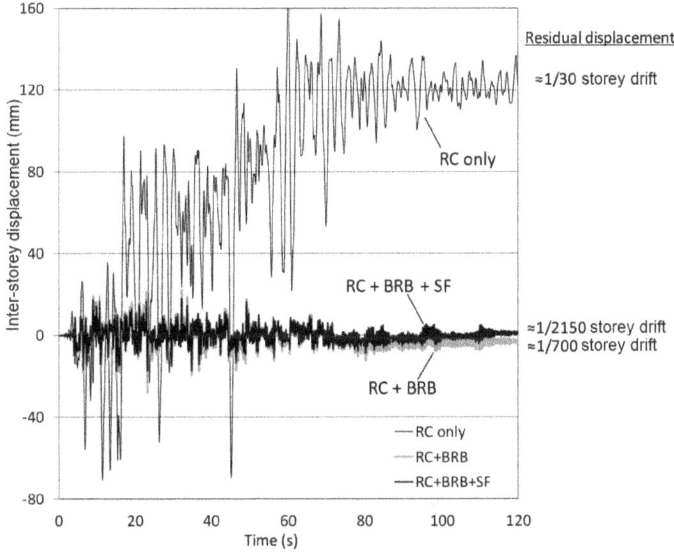

Fig. 4.32 Inter-storey displacement time-history (2nd storey) [64]

supplementary steel frame. In general, the BRBs are effective in controlling the response, but only when the steel frame is included, residual drifts are eliminated.

Figure 4.33 shows the full-scale retrofit subassembly test at Istanbul Technical University to validate the composite response and connection details [70]. The test specimens are identified as:

- R model: Bare RC frame
- RS model: RC frame with concentric steel frame
- RSB model: RC frame with concentric steel frame and BRB

Fig. 4.33 Cyclic loading test for RC retrofit with BRB+ SF (Photo: F. Sutcu)

The RS model was investigated as an intermediate step of the proposed retrofit method, where the test results could help distinguish the effect of the steel frame. Experimental load-displacement hysteresis loops of the cyclic loading tests were compared with the proposed model, and the results are presented in Figure 4.34. The equivalent damping ratio along the amplitudes and cycles have also been calculated, and the results are presented in Figure 4.35. The main observations from the experimental campaign are the following: (a) The addition of the BRB and steel frame improved the structural performance significantly. (b) At the 1/150 retrofit target storey drift, no significant structural damage was observed in the retrofitted specimen. (c) The lateral strength increased by a factor of 9 compared to the bare RC frame and the energy dissipation by a factor of 3 (Figure 4.34 and Figure 4.35). (d) No global or local buckling of the BRB was observed.

Up to the retrofit target storey drift, the steel frame strain measurements were generally at or below the yield level, indicating that the steel frame can remain elastic up to a target retrofit level. This significantly enhances the self-centring properties of the retrofitted frame during a major earthquake. In actual applications, as the BRBs possess fuller and stable hysteretic behaviour under substantially large deformations, the SF can also be designed to remain elastic in a wider drift range considering greater seismic demands or larger target drifts. Strain gauges were also attached to the rebar in the RC frame. Although several indicated yielding, the structural integrity of the RC frame was maintained, with minimal cracking and ductile behaviour observed with controlled distributed plasticity along with the RC members.

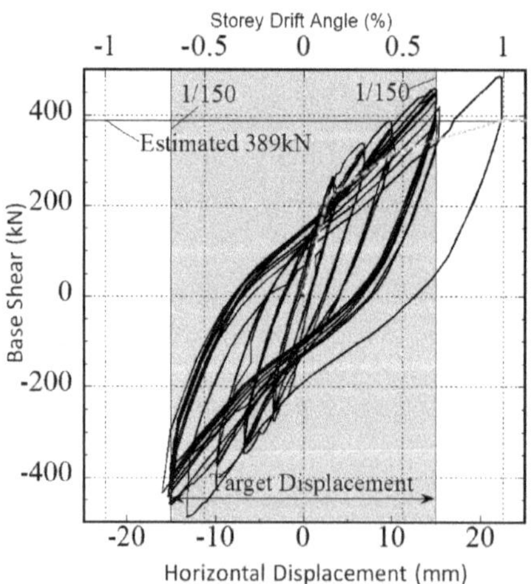

Fig. 4.34 Obtained hysteresis compared with the proposed model [64]

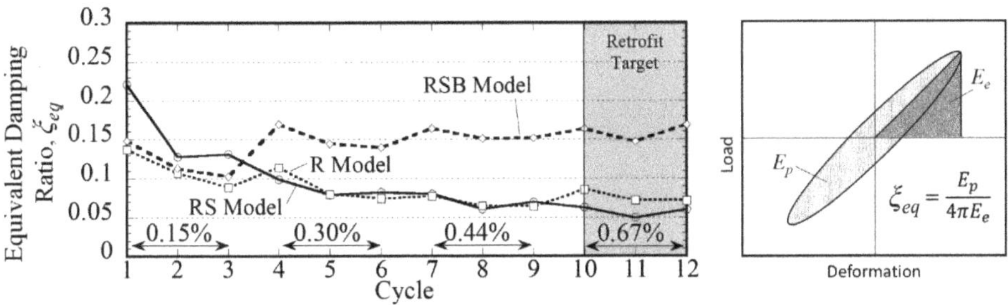

Fig. 4.35 Equivalent damping ratio along the amplitudes and cycles [64]

4.2.3 Retrofit of an 8 Storey RC Building Using Viscous Dampers, Christchurch, New Zealand

The seismic strengthening of the Rehua Building involves an 8-storey ductile reinforced concrete two-way moment-frame located in Christchurch, New Zealand. The building was constructed in the mid-1990s, with Capacity Design and modern ductile reinforcing detailing used in the primary frame members. An assessment of the building identified that storey drifts were likely to be in excess of 5 % for the Design-Basis Earthquake (1/1000-year event), see Figure 4.36.

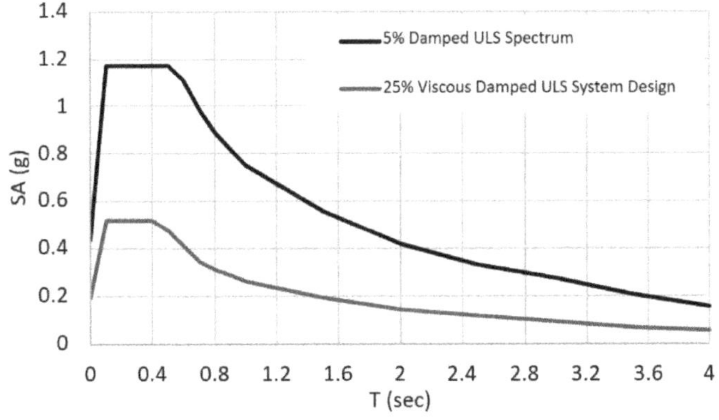

Fig. 4.36 Elastic design spectrum and associated design spectrum with target viscous damping and ductility force-reduction factor (μ_{eff}) applied

4.2.3.1 Objective of the Project

A seismic strengthening scheme using Fluid-Viscous Dampers (FVD) and BRBs was developed to address the excessive drifts, with a retrofit storey drift target of 1.5 % under design loading. This target was set to address potential damage to or loss of the precast hollow core seating as a result of the moment-frame rotations and beam elongation (that occurs with great ductility demands). The introduction of FVDs was also provided to limit floor acceleration increases that would otherwise result from more traditional braced-frame approaches to stiffening the structure.

(a)

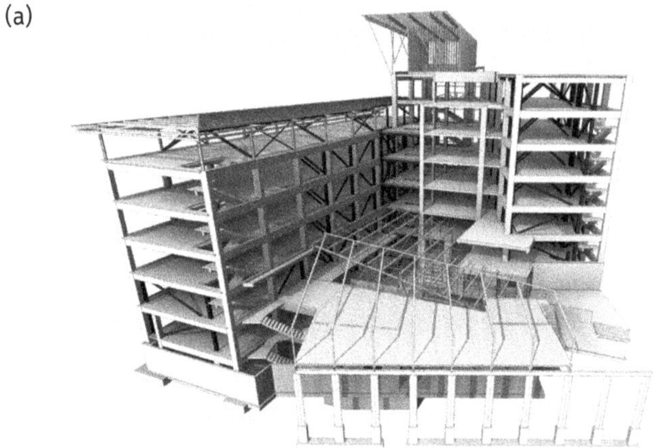

(b)

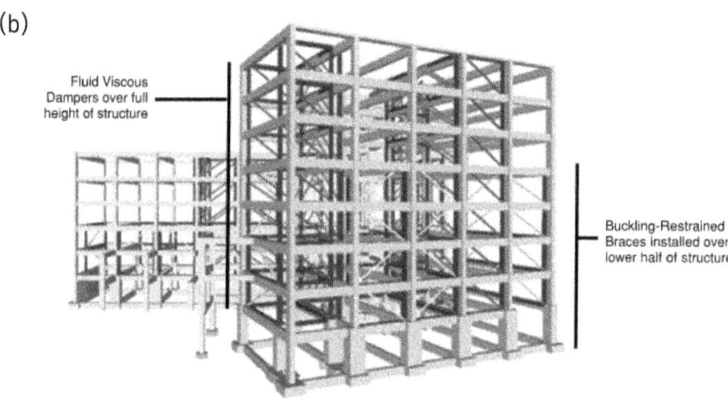

Fig. 4.37 (a) 3D model of the structural frame and (b) NLTHA model

Figure 4.37(a) provides a 3D image of the building complex, including strengthening frames shown in red (a), along with the non-linear time-history analysis model (Figure 4.37 b)) that shows the extend of fluid viscous dampers and buckling-restrained braces over the building elevation.

Retrofit with FVDs is often considered advantageous for limiting additional force demands on the existing columns and foundations due to the device resistance being velocity dependent. However, in practice, this advantage is somewhat reduced in effect due to the axial flexibility of the supporting frame structure. This flexibility leads to the peak FVD restoring force tending to develop only slight out-of-phase to the primary frame structure. In retrofit circumstances, the opportunity to axially stiffen the connecting structure is limited, and designers should be aware of this interaction and the potential reduced efficiency of the dampers.

4.2.3.2 Design and Performance Confirmation

In the example presented here, the reduction in seismic response from 5 % to 1.5 % storey drift was more than could be achieved with FVDs alone, particularly when frame-stiffness

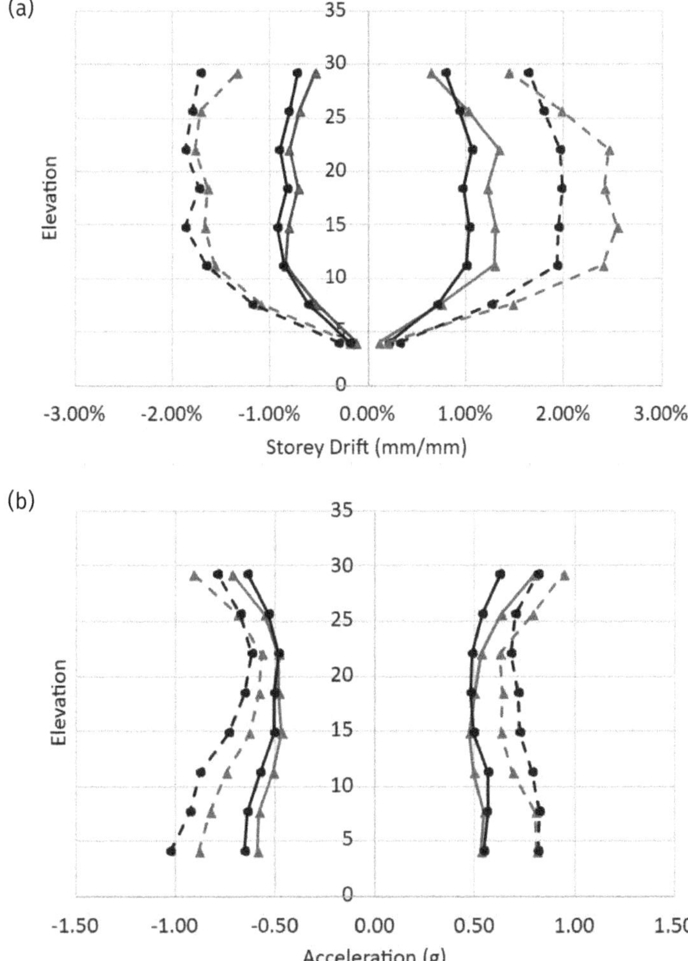

Fig. 4.38 Envelopes for 1/1000 yr (solid) and 1/2000 yr (dashed) earthquake response: (a) inter-storey drifts and (b) peak floor accelerations. Black and red lines denote global X and Y in-plan response directions.

interaction was included. The use of BRBs was, therefore, a means to stiffen the structure over the critical lower floors, to the point that the dampers could effectively reduce the building deformations to meet the strengthening target and control the floor acceleration demands. Figure 4.38 provides summary envelopes of peak storey drift and floor acceleration for both the design-level (ULS) and collapse-limit (CLS) ground motions. It is notable that the target drift was accurately met with the combined retrofit solution while maintaining relatively low floor accelerations.

The FVDs were designed to provide the necessary restoring force contribution and supplemental damping ratio for the 1/1000 year DBE. This was evaluated using a displacement-based approach where the target storey drift was defined by the 1.5% limit noted above. The direct displacement-based approach used is described by [71].

Fig. 4.39 A fluid-viscous damper with extension tube to create a damped frame
(Photo: J. White)

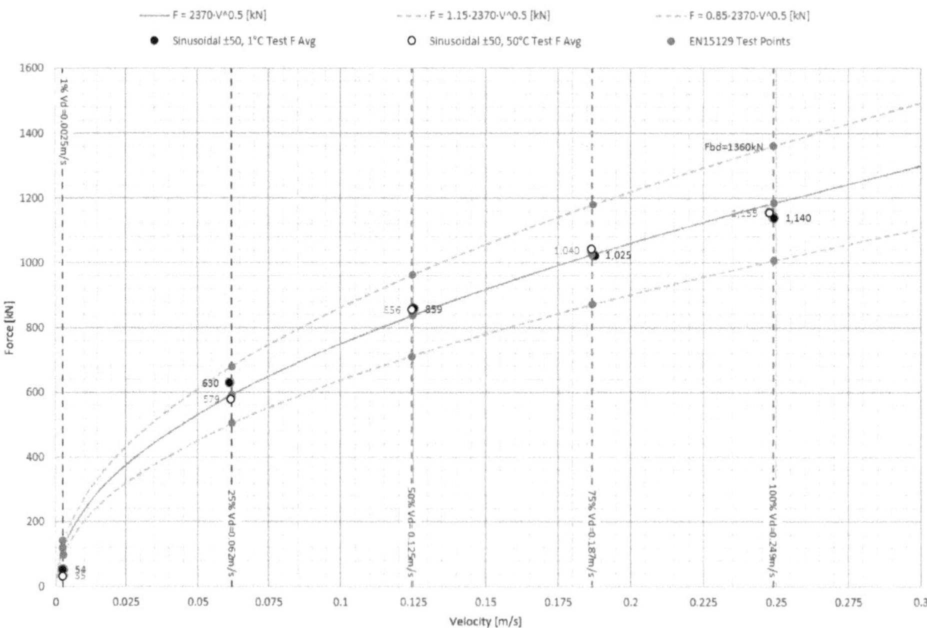

Fig. 4.40 An example of a fluid viscous damper performance curve for a range of activation
velocities from Serviceability to MCE. The red and blue data points reflect testing
at 1°C and 50°C, relative to the target performance (solid curve) and acceptable
margin (dashed curve).

Non-linear time-history analyses were used to verify the acceptable performance of the strengthened building. Suites of records were used to confirm both DBE response and MCE response. The force and stroke capacity for the dampers at each location in the building was specified from the MCE (1/2000 year) event to ensure a level of reserve capacity was present in the system.

Given the significant reduction in building response that was targeted, the dampers in the lower levels of the building had an approximate capacity of 2400 kN. Figure 4.39 is an example of one of these dampers in its final condition.

Specification of the FVDs should refer to the internationally accepted design and testing codes. These will typically allow for an amount of variation from the specified performance. However, the designer should ensure that over the potential range of response velocities, the damper force recorded during prototype or quality control testing does not fall outside the acceptable tolerance. The example of the testing response curve (force vs stroke velocity) shown in Figure 4.40 shows the acceptable margin from the design specification response.

Chapter 5

Post-Earthquake Survey Observations

Post-earthquake survey observations and monitoring can provide important information about the effectiveness of the examined seismic isolated and response-controlled structures. Visual inspection of the buildings is normally conducted after major earthquake events. However, the lack of specialised monitoring systems leads, mostly, to qualitative observations about the efficiency of the seismically isolated/response-controlled structures with limited reliability. Post-earthquake data from monitoring systems in these structures can offer valuable information about the effectiveness of the examined schemes. In this section, two such cases of buildings in Japan are presented.

5.1 Ishinomaki Red Cross Hospital, Seismically Isolated Hospital Building, Ishinomaki, Japan

Ishinomaki Red Cross Hospital is located in Ishinomaki city, Miyagi prefecture, Japan, which is the only hospital designated as a disaster hospital in Ishinomaki medical zone (Figure 5.1). In addition to emergency rescue, the hospital was given the role of accepting and transporting the sick and injured within the disaster area. The footprint and total floor area of the main hospital building are $10{,}173\,\mathrm{m}^2$ and $32{,}486\,\mathrm{m}^2$, respectively. The building has seven storeys above the ground level and one floor at the basement, with a total height of $26.2\,\mathrm{m}$. The design was performed by Nikken Sekkei Co.Ltd. and the contractor was Kajima Corporation. The construction period was from August 2004 to June 2006 [69], [72].

The section and plan views of the main building are shown in Figure 5.2(a) and (b), respectively, where the location of the isolation systems is indicated. Natural Rubber Bearings (NRB) and flat sliding bearings are used as isolation bearings, and U-shape steel dampers were also added (Figure 5.2(b)). The natural period of the structure only with rubber bearings is 5.39 sec, which is reduced to 3.73 sec, including the equivalent stiffness of sliding bearings at a displacement of 490 mm. The total shear force of U-dampers and sliders equals almost 5 % of the total building weight.

During the Great East-Japan Earthquake on March 11th 2011, Ishinomaki city recorded the ground motion of $PGA=633\,\mathrm{cm/sec}^2$, and the isolated layer of the building recorded a maximum displacement of 260 mm in the east-west direction. This is almost half

(a)

(b)

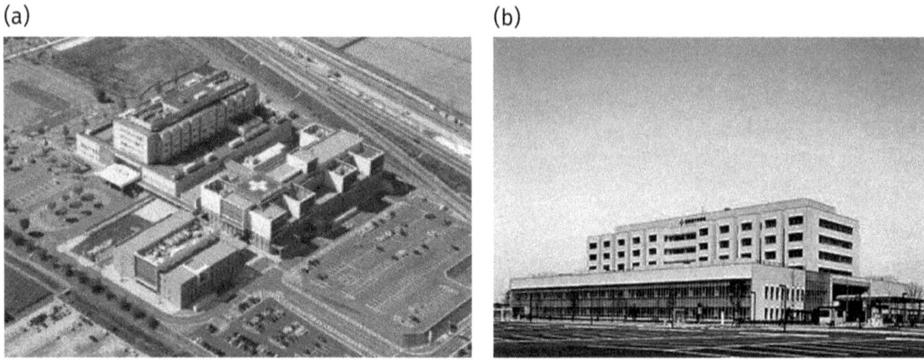

Fig. 5.1   Ishinomaki Red Cross Hospital [72]

(a)

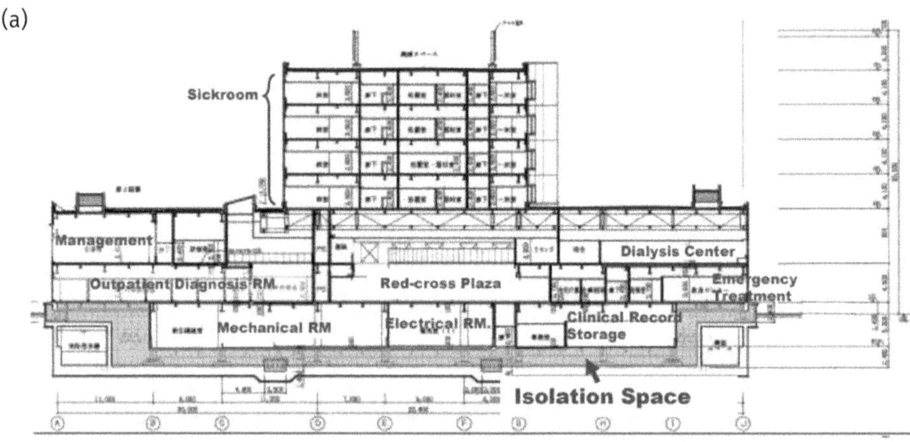

(b)

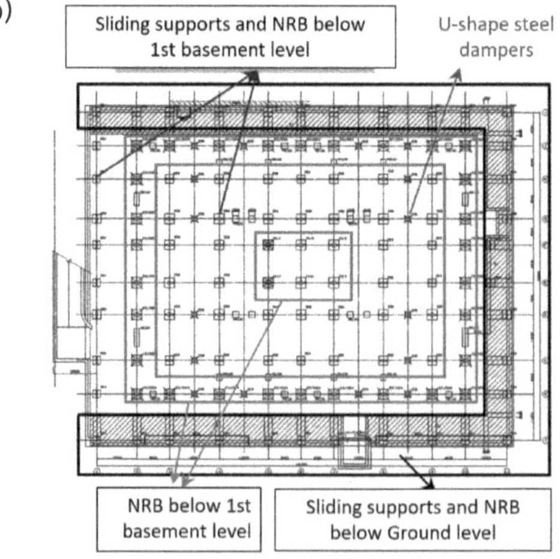

Fig. 5.2   (a) Section and (b) plan view of Ishinomaki Red Cross Hospital main building [72]

of the maximum allowable displacement of 490 mm. The maximum acceleration response at the sixth floor was expected to be reduced to around 150 cm/sec^2. In the whole building, there was no significant damage, including the main medical facilities and machines, while no fatalities or injured people occurred. Bookshelves, folders and documents piled up on the desks fell down, as shown in Figure 5.3, and rescue activities started after the place was cleaned.

The entrance hall on the first floor was rearranged for receiving light-injured people, while the waiting hall was occupied with heavy-injured people (Figure 5.4). Temporary rescue tents were set in front of the main building, where injured people were treated (Figure 5.5).

Fig. 5.3 Fallen documents at office [72]

Fig. 5.4 Open space use after disaster: (a) entrance hall at usual use, (b) entrance hall after the earthquake, (c) waiting hall at usual use and (d) waiting hall after the earthquake [72]

Fig. 5.5   Temporary rescue tents for emergency operations [72]

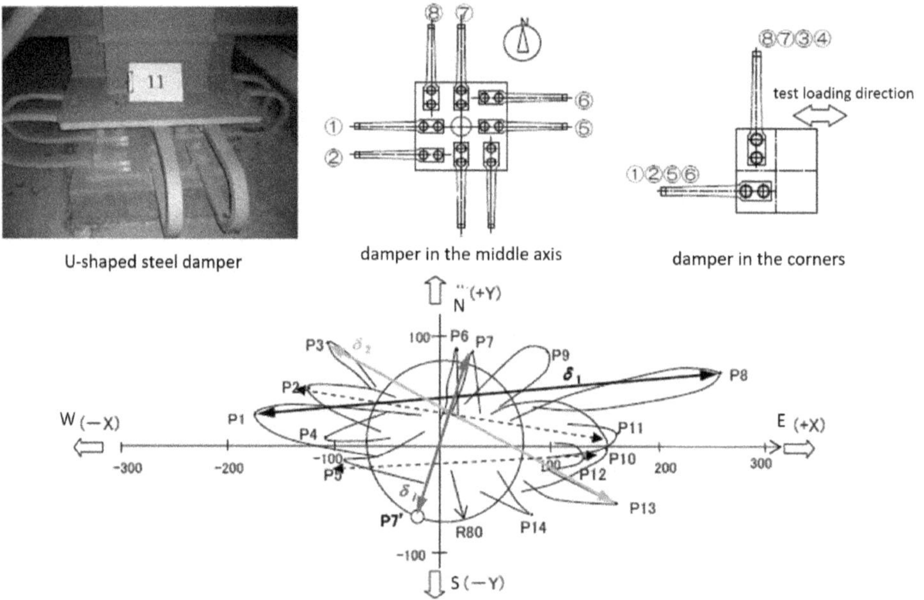

Fig. 5.6   Orbit record at isolation layer [72]

The orbit of the isolated layer record is shown in Figure 5.6. The maximum range of
the movement was +260 mm to -170 mm in the east-west direction and +100 mm to
-60 mm in the north-south direction. No tear or damage was observed in the isolation
bearings. The low-cycle elastoplastic deformation capacities of U-shape steel dampers
were examined. Fatigue tests on the dampers indicate that there is still enough plastic
deformation capacity in each damper. Therefore, it was decided to continue using them
after the earthquake.

5.2 Koriyama Big-Eye Building, High-Rise Building with Viscoelastic Dampers and BRBs, Fukushima, Japan

Koriyama Big-eye building is a steel high-rise building with viscoelastic dampers and BRBs that was subjected to design level ground motions during the 2011 Tohoku Earthquake achieving Immediate Occupancy performance level. Figure 5.7 shows the Koriyama Big-eye building, a 24-storey, 132.6 m high structure located in Fukushima, Japan, that was completed in 2001. Steel moment frames consisting of square hollow section columns and I-section beams are provided in both directions, designed with a storey yield drift ratio limit of approximately 1 %. The damper system includes BRBs as hysteretic dampers in the first seven storeys to control the seismic response and brace type viscoelastic dampers provided from the eighth to twentieth stories for both seismic and wind vibration. The BRBs are designed and manufactured to yield at just 0.13~0.16 % storey drift ratio, which was achieved by using a low yield steel (LY225) and a core with a short plastic length (L_p/L_0 = 0.25~0.3). The energy dissipation system is effective in keeping the primary frame elastic under a design level earthquake (PGV=50 cm/s) and reduces the steel tonnage compared with conventional ductile design. [32]

(a)   (b)

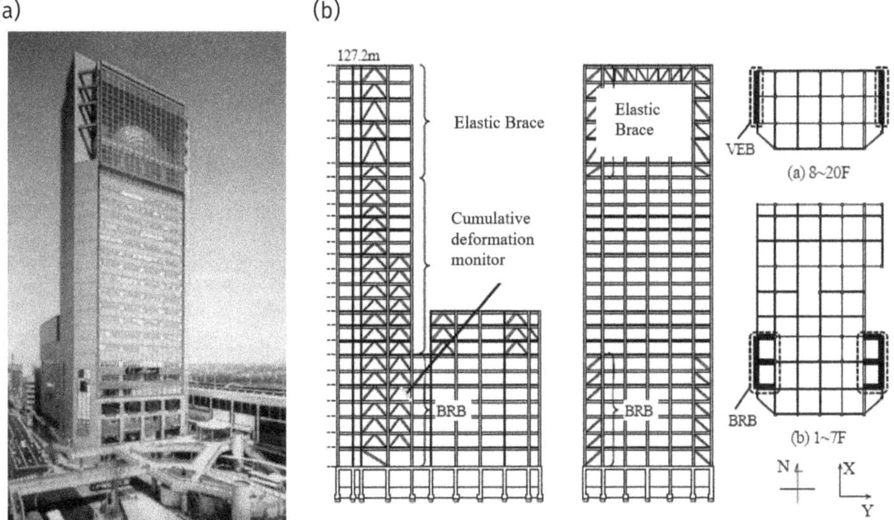

Fig. 5.7   Koriyama Big-Eye building, Fukushima, Japan: (a) photograph and (b) side view drawings (Photo: T. Takeuchi) [56]

One of the main design targets was to keep the primary frame in an elastic range under a design level seismic event. During the design stage, the effect of the use of 2 different damper types was examined through time-history analyses of different damper configurations and combinations under a set of seismic motions. Figure 5.8 shows the comparison of maximum storey displacement results for hysteretic and viscoelastic dampers (Figure 5.8(a)), only with hysteretic dampers (Figure 5.8(b)), and only viscoelastic dampers (Figure 5.8(c)), and the efficiency of the combination of hysteretic and viscoelastic dampers is apparent.

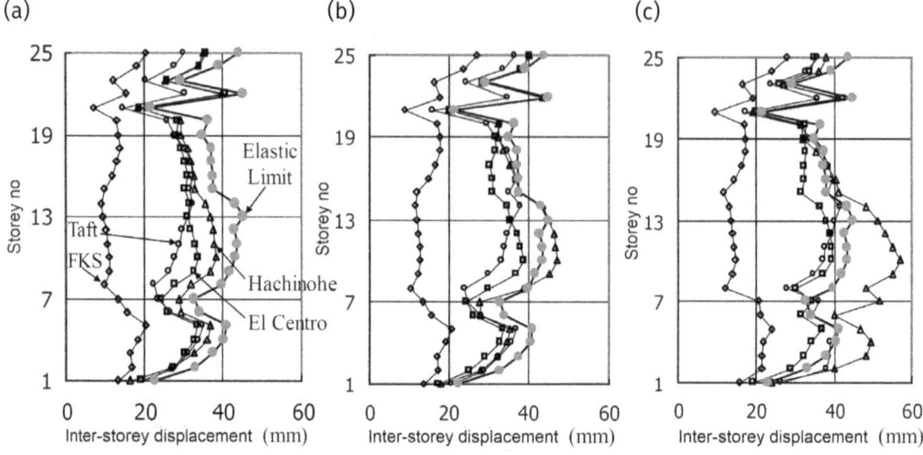

Fig. 5.8 Time-history analyses results (a) with hysteretic and viscoelastic dampers, (b) only with hysteretic dampers and (c) only viscoelastic dampers [56]

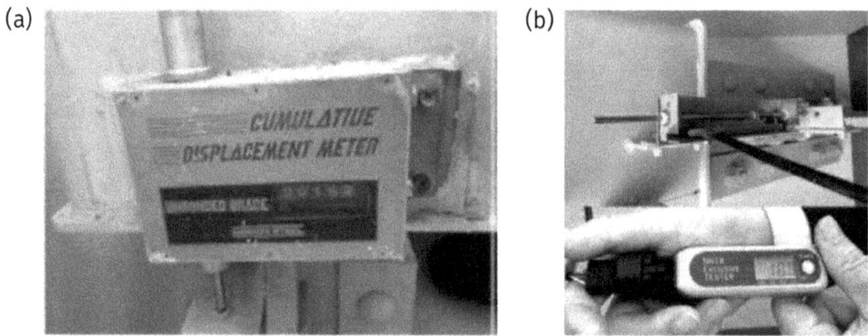

Fig. 5.9 (a, b) Deformation monitoring devices on the Koriyama Big-Eye building (Photo: T. Takeuchi)

During the post-disaster surveys after the 2011 Tohoku Earthquake, significant yielding was recorded at Koriyama Big-Eye Building (234 km from epicentre), in which cumulative and peak deformation recording devices were installed (Figure 5.9). A nearby K-Net station (FKS018) recorded 1.1 g PGA but a spectral acceleration of just 0.08 g (T=3 s, h=5 %). The cumulative deformation measurements and earthquake record were used to calibrate a finite element model, determine the strain time history and estimate the residual cumulative damage index.

The initial visual inspection suggested only a possibility of nominal yielding, a desktop study benchmarked against the deformation readings and indicated a peak ductility demand of $\mu \approx 3.8$ and a cumulative plastic strain of $\sum \varepsilon_p \approx 22\ \% \ (\sum \varepsilon_p / \varepsilon_y \approx 200)$ in the Y direction. During prequalification testing, the supplier had conducted constant amplitude tests to derive the fatigue curve. This was used to confirm that the actual damage index from this event was less than 6 % of the cumulative capacity (in terms of axial plastic deformation), leaving plenty of residual capacity for expected future strong ground motions and justifying the decision to leave all BRBs in place.

This experience indicates several key learning points for post-earthquake inspection:

1) Visual inspection is an unreliable indicator of peak or cumulative ductility demand.

2) Displacement monitoring devices are invaluable and the only reliable means to confirm the ductility demand and reserve capacity in the absence of direct access to the core. It is important to ensure that the monitoring devices are accessible, the building maintenance staff are familiar with the function and operation of the devices, and that periodic inspection and records are maintained, including those after all large earthquakes.

3) Site acceleration records and BRB-specific fatigue curves can be used to provide quantitative justification of the residual capacity.

4) Well-designed BRBs can withstand multiple severe earthquakes, and replacement is not necessarily required, even following significant yielding.

5) No damages were observed in VE dampers.

References

[1]   N. Anwar, T. H. Aung, and F. Najam, 'From Prescription to Resilience: Innovations in Seismic Design Philosophy.', *Technology*, vol. 8, 2016.

[2]   REDi, 'REDiTM rating system: Resilience-based earthquake design initiative for the next generation of buildings. Arup Publications, 2013.

[3]   Department of Homeland Security (DHS), 'National Infrastructure Protection Plan'. Washington D.C.: Department of Homeland Security, 2009.

[4]   M. Melkumyan, 'The behavior of retrofitted buildings during earthquakes: New technologies', in *Building safer cities: the future of disaster risk*, Washington, D.C: World Bank, 2003, pp. 293–299.

[5]   https://commons.wikimedia.org/wiki/File:Oakland_City_Hall_(Oakland,_CA)_2. JPG

[6]   https://commons.wikimedia.org/w/index.php?curid=5658910

[7]   G. Mylonakis and G. Gazetas, 'Seismic soil-structure interaction: beneficial or detrimental?', *Journal of Earthquake Engineering*, vol. 4, no. 3, pp. 277–301, Jul. 2000. https://doi.org/10.1080/13632460009350372

[8]   J. Kelly, 'Shake table tests of long period isolation system for nuclear facilities at soft soil sites', University of California at Berkeley, UBC/EERC-91/03, 1991.

[9]   C. S. Tsai, C.-S. Chen, and B.-J. Chen, 'Effects of unbounded media on seismic responses of FPS-isolated structures', *Structural Control & Health Monitoring*, vol. 11, no. 1, pp. 1–20, Jan. 2004. https://doi.org/10.1002/stc.28

[10] C. C. Spyrakos, Ch. A. Maniatakis, and I. A. Koutromanos, 'Soil–structure interaction effects on base-isolated buildings founded on soil stratum', *Engineering Structures*, vol. 31, no. 3, pp. 729–737, Mar. 2009. https://doi.org/10.1016/j. engstruct.2008.10.012

[11] G. Manolis and AA. Markou, 'A distributed-mass structural system for soil-structure-interaction and base isolation studies', Special Issue honoring Professor Anthony N. Kounadis on the occasion of his 75th birthday. *Archive of Applied Mechanics*, vol. 82, pp. 1513–1529, 2012. https://doi.org/10.1007/s00419-012-0659-8

[12] C. Giarlelis, J. Keen, E. Lamprinou, V. Martin, and G. Poulios, 'The seismic isolated Stavros Niarchos Foundation Cultural Center in Athens (SNFCC)', *Soil Dynamics and Earthquake Engineering*, vol. 114, pp. 534–547, Nov. 2018. https://doi. org/10.1016/j.soildyn.2018.05.011

[13] T. Tomizawa *et al.*, 'Vibration test in a Building named "Chisuikan" using Three-dimensional Seismic Isolation System', presented at the 15WCEE, Lisbon, Portugal, 2012.

[14] https://commons.wikimedia.org/wiki/File:GERB_spring_with_damper.jpg

[15] Bridgestone, 'Seismic Isolation Product Line-up: High Damping Rubber Bearing, Lead Rubber Bearing, Natural Rubber Bearing and Elastic Sliding Bearing'. Bridgestone Corporation, 2017.

[16] P. M. Calvi and G. M. Calvi, 'Historical development of friction-based seismic isolation systems', *Soil Dynamics and Earthquake Engineering*, vol. 106, pp. 14–30, Mar. 2018. https://doi.org/10.1016/j.soildyn.2017.12.003

[17] S. Barone, G. M. Calvi, and A. Pavese, 'Experimental dynamic response of spherical friction-based isolation devices', *Journal of Earthquake Engineering*, vol. 23, no. 9, pp. 1465–1484, Oct. 2019. https://doi.org/10.1080/13632469.2017.1387201

[18] D. Cardone, G. Gesualdi, and P. Brancato, 'Restoring capability of friction pendulum seismic isolation systems', *Bulletin of Earthquake Engineering*, vol. 13, no. 8, pp. 2449–2480, Aug. 2015. https://doi.org/10.1007/s10518-014-9719-5

[19] A. Tsiavos, T. Markic, D. Schlatter, and B. Stojadinovic, 'Shaking table investigation of inelastic deformation demand for a structure isolated using friction-pendulum sliding bearings', p. 12 p., 2021. https://doi.org/10.3929/ETHZ-B-000474263

[20] A. Pavese, M. Furinghetti, and C. Casarotti, 'Investigation of the Consequences of Mounting Laying Defects for Curved Surface Slider Devices under General Seismic Input', *Journal of Earthquake Engineering*, vol. 23, no. 3, pp. 377–403, Mar. 2019. https://doi.org/10.1080/13632469.2017.1323046

[21] A. Mokha, M. C. Constantinou, A. M. Reinhorn, and V. A. Zayas, 'Experimental Study of Friction-Pendulum Isolation System', *Journal of Structural Engineering*, vol. 117, no. 4, pp. 1201–1217, Apr. 1991. https://doi.org/10.1061/(ASCE)0733-9445(1991)117:4(1201).

[22] P. Tsopelas, C. Constantinou, Y. S. Kim, and S. Okamoto, 'Experimental Study of FPS System in Bridge Seismic Isolation', *Earthquake Engineering and Structural Dynamics.*, vol. 25, no. 1, pp. 65–78, 1996. https://doi.org/10.1002/(SICI)1096-9845(199601)25:1<65::AID-EQE536>3.0.CO;2-A

[23] F. Naeim and J. M. Kelly, *Design of seismic isolated structures: from theory to practice*. New York: John Wiley, 1999.

[24] Nippon Steel Engineering, 'Nippon Steel-Spherical Sliding Bearing Catalogue'. Nippon Steel, 2020.

[25] THK CO., LTD., 'THK Base Isolation Catalog – Technical Book'. THK CO., LTD.

[26] Nippon Steel & Sumikin Engineering, 'NS-U (U-Shaped Steel Damper)'. Nippon Steel & Sumikin Engineering, 2020.

[27] T. Takeuchi, *Design of Seismic Isolation and Response Control*. AIJ Kanto Press, 2007.

[28] D. Taylor and M. Constantinou, 'Fluid Dampers for Applications of Seismic Energy Dissipation and Seismic Isolation'. Taylor Devices Inc., 2000.

[29] V. A. Zayas, S. S. Low, and S. A. Mahin, 'A Simple Pendulum Technique for Achieving Seismic Isolation', *Earthquake Spectra*, vol. 6, no. 2, pp. 317–333, May 1990. https://doi.org/10.1193/1.1585573

[30] I. G. Buckle, M. Constantinou, M. Dicleli, and H. Ghasemi, 'Seismic isolation of highway bridges', University of Buffalo, NY, Research Report MCEER-06-SP07.

[31] C. Giarlelis, C. Kostikas, E. Lamprinou, and M. Dalakiouridou, 'Dynamic behavior of a seismic isolated structure in Greece', Beijing, China, 2008.

[32] Japan Society of Seismic Isolation (JSSI), 'JSSI Manual: Design and Construction Manual for Passively Controlled Buildings'. 2003.

[33] JFE Civil Engineering & Construction Corp., 'JFE Vibration Control Column Catalogue'. 2019.

[34] T. Sano, K. Shirai, Y. Suzui, and Y. Utsumi, 'Loading tests of a brace-type multi-unit friction damper using coned disc springs and numerical assessment of its seismic response control effects', *Bulletin of Earthquake Engineering*, vol. 17, no. 9, pp. 5365–5391, Sep. 2019. https://doi.org/10.1007/s10518-019-00671-8

[35] T. Takeuchi, R. Matsui, and S. Mihara, 'Out-of-plane stability assessment of buckling-restrained braces including connections with chevron configuration', *Earthquake Engineering and Structural Dynamics*, vol. 45, no. 12, pp. 1895–1917, Oct. 2016. https://doi.org/10.1002/eqe.2724

[36] OILES Corporation, 'Viscous Wall Damper', 2020. https://www.oiles.co.jp/en/menshin/building/seishin/products/vwd/

[37] A. Di Cesare, F. C. Ponzo, D. Nigro, M. Dolce, and C. Moroni, 'Experimental and numerical behaviour of hysteretic and visco-recentring energy dissipating bracing systems', *Bulletin of Earthquake Engineering*, vol. 10, no. 5, pp. 1585–1607, Oct. 2012. https://doi.org/10.1007/s10518-012-9363-x

[38] E. Tubaldi, L. Gioiella, F. Scozzese, L. Ragni, and A. Dall'Asta, 'A Design Method for Viscous Dampers Connecting Adjacent Structures', *Frontiers in Built Environment*, vol. 6, p. 25, Mar. 2020. https://doi.org/10.3389/fbuil.2020.00025

[39] M. Dolce, D. Cardone, and R. Marnetto, 'Implementation and testing of passive control devices based on shape memory alloys', *Earthq. Eng. Struct. Dyn.*, vol. 29, no. 7, pp. 945–968, 2000. https://onlinelibrary.wiley.com/doi/10.1002/1096-9845(200007)29:7%3C945::AID-EQE958%3E3.0.CO;2-%23

[40] FIP Industriale, 'Anti-Seismic Devices'. 2016.

[41] CSN EN 15129, 'Anti-seismic devices.' Brussels: European Committee for Standardisation, 2009.

[42] European Committee for Standardization (CEN), *Eurocode 8: Design of Structures for Earthquake Resistance-Part 1: General Rules, Seismic Actions and Rules for Buildings*. 2004, p. 229.

[43] American Society of Civil Engineers, *ASCE/SEI 7-16: Minimum Design Loads for Buildings and Other Structures*. 2016. https://doi.org/10.1061/9780784414248

[44] *Japanese Seismic Code: The Notification and Commentary on the Structural Calculation Procedures for Building with Seismic Isolation*. 2000.

[45] NZSEE, 'Guideline for the Design of Seismic Isolation Systems for Buildings'. Jun. 2019.

[46] CFE: Federal Electricity Commission, 'Manual of Civil Structures in Mexico: Seismic Design'. Cuernavaca, Morelos, Mexico, 2015.

[47] Turkish Disaster and Emergency Management Authority (AFAD), 'Turkish Building Seismic Code'. 2018.

[48] Tecno K Giunti, 'Seismic Joint Covers'. Tecno K Giunti S.r.I., 2020.

[49] CFE: Federal Electricity Commission, 'MDS-CFE: Manual de Diseño de Obras Civiles (Diseño por Sismo)'. Cuernavaca, Morelos, Mexico, 1993.

[50] J. M. Jara, E. Madrigal, M. Jara, and B. A. Olmos, 'Seismic source effects on the vulnerability of an irregular isolated bridge', *Engineering Structures*, vol. 56, pp. 105–115, Nov. 2013. https://doi.org/10.1016/j.engstruct.2013.04.022

[51] J. M. Jara, M. Jara, H. Hernández, and B. A. Olmos, 'Use of sliding multirotational devices of an irregular bridge in a zone of high seismicity', *KSCE Journal of Civil Engineering*, vol. 17, no. 1, pp. 122–132, Jan. 2013. https://doi.org/10.1007/s12205-013-1063-9.

[52] A. Ghobarah, A. Biddah, and M. Mahgoub, 'Rehabilitation of Reinforced Concrete Columns using Corrugated Steel Jacketing', *Journal of Earthquake Engineering*, vol. 01, pp. 651–673, 1997. http://doi.org/10.1080/13632469708962382

[53] R. Watson, 'EradiQuake Isolation bearing System', *http://www.roadauthority.com/Product/Details/3312*, May 25, 2020.

[54] RJ Watson, Inc., 'ERADIQUAKE: Isolation & Force Control Bearing Devices – Innovation by Design'. RJ Watson, Inc. Bridge & Structural Engineered Systems, 2019.

[55] M. D. Symans *et al.*, 'Energy Dissipation Systems for Seismic Applications: Current Practice and Recent Developments', *Journal of Structural Engineering*, vol. 134, no. 1, pp. 3–21, Jan. 2008. https://doi.org/10.1061/(ASCE)0733-9445(2008)134:1(3).

[56] T. Takeuchi and A. Wada, *Buckling-restrained Braces and Applications*. Japanese Society of Seismic Isolation Press, 2017.

[57] T. Takeuchi, 'Structural design with seismic energy-dissipation concept', in *IABSE Conference 2015*, Geneva, Switzerland, Sep. 2015, p. 2157. https://doi.org/10.2749/222137815815773747

[58] American Society of Civil Engineers, *ASCE/SEI 41-17: Seismic Evaluation and Retrofit of Existing Buildings*. 2017.

[59] M. Erdik, Ö. Ülker, B. Şadan, and C. Tüzün, 'Seismic isolation code developments and significant applications in Turkey', *Soil Dynamics and Earthquake Engineering*, vol. 115, pp. 413–437, Dec. 2018. https://doi.org/10.1016/j.soildyn.2018.09.009

[60] C. Giarlelis, D. Koufalis, and C. Repapis, 'Seismic Isolation: An Effective Technique for the Seismic Retrofitting of a Reinforced Concrete Building', *Structural Engineering International*, vol. 30, no. 1, pp. 43–52, Jan. 2020. https://doi.org/10.1080/10168664.2019.1678449

[61] European Committee for Standardization (CEN), *Eurocode 8: Design of Structures for Earthquake Resistance-Part 3: Assessment and Retrofitting of Buildings*. 2005, p. 81.

[62] The Japan Disaster Prevention Association (JDPA), 'Recommendation for Seismic Retrofit for Reinforced Concrete Buildings'. 1989.

[63] T. Takeuchi, K. Yasuda, and M. Iwata, 'Studies on Integrated Building Façade Engineering with High-Performance Structural Elements', IABSE Symposium Budapest, 2006, pp. 442–443. https://doi.org/10.2749/222137806796185526

[64] F. Sutcu, T. Takeuchi, and R. Matsui, 'Seismic retrofit design method for RC buildings using buckling-restrained braces and steel frames', *Journal of Constructional Steel Research*, vol. 101, pp. 304–313, Oct. 2014. https://doi.org/10.1016/j.jcsr.2014.05.023

[65] JBDPA (Japan Building Disaster Prevention Association), 'Standard for Seismic Diagnosis of Existing Reinforced Concrete Structures'. 2001.

[66] FEMA 273, 'NEHRP Guidelines for the Seismic Rehabilitation of Buildings'. Prepared for FEMA by the Applied Technology Council and the Building Seismic Safety Council. Washington, DC, 1997.

[67] FEMA 356, 'Prestandard and Commentary for the Seismic Rehabilitation of Buildings'. Prepared for FEMA by the American Society of Civil Engineers. Washington, DC, 2000.

[68] Applied Technology Council, 'ATC 40 Report: Seismic Evaluation and Retrofit of Concrete Buildings', 1996.

[69] N. Kawamura and Konishi, 'Evaluation of the Fatigue Life of U-shaped Steel Dampers after Extreme Earthquake Loading', Sendai, Japan, Sep. 2013.

[70] F. Sutcu, A. Bal, K. Fujishita, R. Matsui, O. C. Celik, and T. Takeuchi, 'Experimental and analytical studies of sub-standard RC frames retrofitted with buckling-restrained braces and steel frames', *Bulletin of Earthquake Engineering*, vol. 18, no. 5, pp. 2389–2410, Mar. 2020. https://doi.org/10.1007/s10518-020-00785-4

[71] M. J. N. Priestley, G. M. Calvi, and M. J. Kowalsky, *Displacement-based seismic design of structures*. Pavia, Italy: IUSS Press : Distributed by Fondazione EUCENTRE, 2007.

[72] T. Someya, 'Seismically Isolated Hospital Offers Ray of Hope in Disaster – Ishinomaki Red Cross Hospital', Sendai, Japan, Sep. 2013.